GEOGRAPHICAL KNOWLEDGE AND IMPERIAL CULTURE IN THE EARLY MODERN OTTOMAN EMPIRE

Transculturalisms, 1400–1700

Series Editors:
Mihoko Suzuki, University of Miami, USA,
Ann Rosalind Jones, Smith College, USA, and
Jyotsna Singh, Michigan State University, USA

This series presents studies of the early modern contacts and exchanges among the states, polities and entrepreneurial organizations of Europe; Asia, including the Levant and East India/Indies; Africa; and the Americas. Books will investigate travelers, merchants and cultural inventors, including explorers, mapmakers, artists and writers, as they operated in political, mercantile, sexual and linguistic economies. We encourage authors to reflect on their own methodologies in relation to issues and theories relevant to the study of transculturism/translation and transnationalism. We are particularly interested in work on and from the perspective of the Asians, Africans, and Americans involved in these interactions, and on such topics as:

- Material exchanges, including textiles, paper and printing, and technologies of knowledge
- Movements of bodies: embassies, voyagers, piracy, enslavement
- Travel writing: its purposes, practices, forms and effects on writing in other genres
- Belief systems: religions, philosophies, sciences
- Translations: verbal, artistic, philosophical
- Forms of transnational violence and its representations.

Also in this series:

The Turk and Islam in the Western Eye, 1450–1750
Visual Imagery before Orientalism
Edited by James G. Harper

Ottoman Women Builders
The Architectural Patronage of Hadice Turhan Sultan
Lucienne Thys-Senocak

Architecture and Hagiography in the Ottoman Empire
The Politics of Bektashi Shrines in the Classical Age
Zeynep Yürekli

Geographical Knowledge and Imperial Culture in the Early Modern Ottoman Empire

PINAR EMİRALİOĞLU
Sam Houston State University, USA

Routledge
Taylor & Francis Group
LONDON AND NEW YORK

First published 2014 by Ashgate Publishing

Published 2016 by Routledge
2 Park Square, Milton Park, Abingdon, Oxfordshire OX14 4RN
711 Third Avenue, New York, NY 10017, USA

First issued in paperback 2016

Routledge is an imprint of the Taylor & Francis Group, an informa business

British Library Cataloguing in Publication Data
A catalogue record for this book is available from the British Library

The Library of Congress has cataloged the printed edition as follows:
Emiralioğlu, M. Pınar.
Geographical knowledge and imperial culture in the early modern Ottoman Empire / by Pınar Emiralioğlu.
pages cm—(Transculturalisms, 1400–1700)
Includes bibliographical references and index.
ISBN 978-1-4724-1533-2 (hardcover: alk. paper)
1. Turkey—Poltics and government—16th century. 2. Turkey—Politics and government—17th century. 3. Geography—Turkey—History. 4. Geography—Study and teaching—Turkey—History. 5. Cartography—Turkey—History. 6. Imperialism—Social aspects—Turkey. I. Title.
DR486.E45 2014
956.1'0153—dc23

2013026670

ISBN 13: 978-1-138-24754-3 (pbk)
ISBN 13: 978-1-4724-1533-2 (hbk)

For my parents
Gül & Kemal Emiralioğlu

Contents

List of Figures

List of Plates

List of Maps

List of Maps

List of Abbreviations

BSOAS	*Bulletin of the School of Oriental and African Studies*
EI¹	*The Encyclopedia of Islam*, edited by M. Th. Houtsma, T. W. Arnold, R. Basset and R. Hartmann. First Edition. Leiden: E. J. Brill, 1913–1936.
EI²	*The Encyclopedia of Islam*, edited by P. Bearman, Th.Bianquis, C. E. Bosworth, E. van Donzel and W. P. Heinrichs. Second Edition. Leiden: E. J. Brill, 1954–2005.
EJOS	*Electronic Journal of Oriental Studies*
IGAIW	Institut für Geschichte der Arabisch-Islamischen Wissenschaften
JESHO	*Journal of the Economic and Social History of the Orient*
TKS	The Library of the Topkapı Palace Museum
TTK	Türk Tarih Kurumu
ZDMG	*Zeitschrift der Deutschen Morgenländischen Gesellschaft N. F.*

Note on Transliteration, Translation, and Pronunciation

In order to make reading easier for the non-specialist, I have opted for a simplified version of the *International Journal of Middle Eastern Studies* transliteration system for Arabic and Persian terms and personal names. I have avoided most diacritical marks that are not on an English keyboard. I have used the following marks to indicate *hemze* (') and *'ayn* ('). For example, the word "marvels" is transliterated as *'Aja'ib* in Arabic, the word "respected" is transliterated as *mu'tabar* in Persian.

I have rendered Ottoman Turkish personal names, book titles, and quotations throughout the book by means of modern Turkish orthography. I have chosen to use terms familiar to English speakers in their accepted Anglicized version, hence ghazi, mufti, pasha, and vizier. Place names follow *Merriam-Webster's Geographical Dictionary*, 3rd edition (1997). Unless otherwise noted, all translations are mine.

Some Turkish letters are not transliterated in English and some are pronounced differently from their English counterparts:

C,c: "j" as in joy

Ç,ç: "ch" as in church

ğ: unvocalized, lengthens preceding vowel

I,ı: "e" in women

İ,i: "i" in bit

Ö,ö: "eu" as in French *deux*

Ş,ş: "sh" as in shade

Ü,ü: "ü" as in German über

Acknowledgments

In many ways, the story of this book starts at Bilkent University in Ankara in the 1990s when Halil İnalcık, Oktay Özel, and Akşin Somel introduced me and fellow graduate students to the uncharted territories of Ottoman history and taught us to be patient with Ottoman Turkish. At Bilkent, I had the great advantage of finding great friends and colleagues, namely, Özlem Çaykent and Ali Yaycıoğlu who all contributed to my training in the historian's craft with tremendous intellectual support and friendship.

I owe much to Cornell H. Fleischer, who encouraged me to continue my graduate studies at the University of Chicago. He provided invaluable support and advice throughout my entire stay in Chicago and gave me the trust and the freedom to pursue this project in the way I wanted. Robert Dankoff generously offered his tireless patience and philological guidance. In Chicago, I found a large group of Ottoman history students and friends who provided a fascinating learning environment, and threaded strong ties of friendship all over the town from the corridors of the International House of the University of Chicago to the stacks of Regenstein Library. Among them, I am especially indebted to Betül Başaran, Side Emre, Abigail Jacobson, and Nükhet Varlik.

The research for this book was made possible by grants and fellowships from the Andrew Mellon Foundation and Turkish Studies Research Travel Grant of the University of Chicago. The final stages of my archival research also overlapped with my employment in the Department of History at Istanbul Bilgi University. Tremendous thanks go to Mete Tunçay. who was a real mentor and devoted educator. He guided me thoroughly in my attempts to be a better teacher and asked the questions that transformed my dissertation into a trans-national project. Special thanks are also due to Christoph K. Neumann, who read my dissertation and offered his sharp and ample comments and suggestions.

I am also indebted to the staff of numerous archives and libraries that facilitated my research: Süleymaniye Library in Istanbul, Topkapı Palace Library, Bayerische Staats Bibliothek of Munich, and Istanbul University Library. Special thanks go to Esra Müyesseroğlu of the Topkapı Palace Library and Yasemen Akçay of the Istanbul University Library for their generous help in providing the most crucial images for this book.

The story of this book concludes at the University of Pittsburgh. My colleagues at the Department of History helped me reformulate my questions and opened my eyes to the opportunities of trans-national history. They read various drafts of my manuscript and gave invaluable feedback. I am particularly grateful to Bill Chase, Laura Gotzkowitz, Van Beck Hall, Diego Holstein, Pat Manning, and Jonathan Scott. The University of Pittsburgh also provided me with generous institutional support, including two travel grants from the Center for Russian and

Eastern European Studies and World History Center, and financial assistance for the publication of this book by the Richard D. and Mary Jane Edwards Endowed Publication Fund. Pitt and Pittsburgh were also very generous in giving me dear friends, some of whom I should mention here. My neighbor Deborah Chiappini helped me transform my old apartment into a beautiful home. Schuyler Hernstrom read various drafts of this book, gave feedback, prepared figures for the book, and reminded me that there is a universe beyond the Ottomans. Christine Hucko copyedited earlier versions of this work and gave me the final push to complete this study. Christian Gerlach, Vincent Leung, Tony Novasel, Pernille Røge, Rebecca Shumway, Gregor Thum, Ebru Türker, and Molly Warsh all contributed to this study with their support, encouragement, and friendship.

Esteemed colleagues were able to give invaluable feedback. These include the anonymous readers and friends. I am particularly indebted to Thomas D. Goodrich and Svat Soucek. Without their works, this study may not have been possible. I would also like to thank my editor, Erika Gaffney, and assistant editor Patrick Smith for their wise advice, understanding, and attention to detail and the copyeditor Laura Kopp for her careful reading of the manuscript. All the remaining errors in the book are mine.

I could not have completed this book without the love and support of my family and my furry companion Fındık. Over the years, the unconditional love and support of my brother Sinan Emiralioğlu and my parents, Gül and Kemal Emiralioğlu, has made everything possible. I am grateful for all that they have done for me, and it is to them that I dedicate this work.

Map 1 The Ottoman Empire, 1300–1512, adapted from *An Economic and Social History of the Ottoman Empire, Vol. 1: 1300–1600*, ed. Halil İnalcık and Donald Quataert (Cambridge: Cambridge University Press, 1994). Reprinted with the permission of Cambridge University Press.

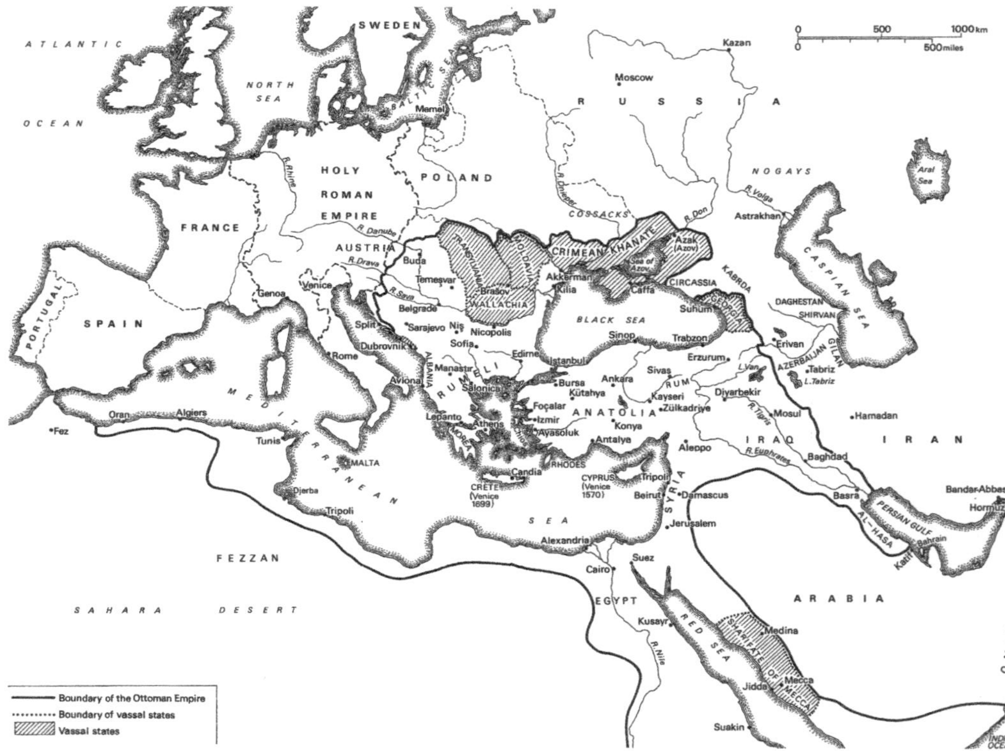

Map 2 The Ottoman Empire, ca. 1550, adapted from *An Economic and Social History of the Ottoman Empire, Vol. 1: 1300–1600*, ed. Halil İnalcık and Donald Quataert (Cambridge: Cambridge University Press, 1994). Reprinted with the permission of Cambridge University Press.

Introduction
Eye of the World: Textual and Visual Repertoires of the Sixteenth-Century Ottoman Empire

Piri Reis entered Selim's tent in Cairo. The Ottoman sultan was sitting on his throne, flanked by servants on both sides. The tent's interior was sumptuously decorated to resemble the sultan's court at the Topkapı Palace in Constantinople.[1] Colorful carpets lined the floor and walls. Inside it was perfectly quiet and still. Piri Reis had never met the sultan before, who was called "Selim the Inexorable" in Constantinople, earning his name with an indomitable will and legendary temper. Piri Reis had heard stories about the sultan cutting off the heads of his rivals and those of crocodiles with the same coldness.

Standing in the tent, Piri Reis remembered his days as a pirate in the Mediterranean. He remembered his cherished independence, free from the authority of any sultan or king. But times had changed. He and his uncle had had to enter the service of the Ottomans two decades earlier. The Ottoman sultans were getting stronger and stronger in the Mediterranean. Pirates or corsairs had to enter their service to survive in this sea.

Piri Reis was holding tightly to a long sheepskin scroll. His future depended on this scroll. He had heard that Selim valued artists and showed his appreciation by bestowing favors upon them. The former pirate considered himself an artist. His medium was navigation. He was born on the sea and spent all his life there doing battle with enemies and fickle weather. He learned how to deal with both, and in the meantime drew charts. He delighted in how the red, blue, and green inks flowed onto the sheepskin, creating beautiful outlines of the islands and the coasts. His uncle always relied heavily on his charts during their ventures. Sequestered in his hometown, Piri Reis had dedicated his last few years to preparing a world map. He knew that Selim's grandfather Mehmed collected maps, but he did not know how Selim would react.

Piri Reis cautiously approached the sultan, leaning forward and keeping his eyes on the intricate kilim underfoot. Without a word, he handed the scroll to one of Selim's servants. Selim took the scroll and opened it. His movements were careful and slow, as if a single misstep would upset the balance of his whole empire.

---

[1] In this book, the names Constantinople and Istanbul will be used interchangeably, as was done by the Ottomans. After they conquered the city in 1453, the Ottomans continued to call it "Kostantaniyye" while occasionally the terms Istanbul or İslambol (full of Islam) appeared on official documents and chronicles as well. The official change of name to Istanbul took place after the foundation of the Turkish Republic in 1923.

His eyes traveled across the scroll from left to right. He did not say a word for what seemed to Piri Reis like hours. Finally Selim's harsh voice filled the tent.

"So, is this the so-called West Indies?"

Piri Reis replied without raising his eyes: "Yes, my Sultan, I copied that part from the charts of sailors who were with Columbus."

"One day every island on this sea shall be part of my protected domain," murmured Selim as he gazed over the Mediterranean.

Fifteen minutes later, Piri Reis left the tent as a sea captain in the imperial fleet in Alexandria and the official cartographer of the Ottoman court. As he stepped out into the cold February weather, he was already planning his future project in his mind—a navigational guide—and exhaling with great relief.

The actors of this story, Piri Reis and the Ottoman Sultan Selim I, in fact met during the sultan's campaign in Egypt in the winter of 1517. Piri Reis, a former pirate who entered the Ottoman service at the end of the fifteenth century, presented the sultan with his world map during this meeting. There is no record of whether Selim, in return, granted him favors or not. But Piri Reis continued to serve in the Ottoman navy as a captain under Selim I and Süleyman I. In 1547, the captain was appointed commander of the Ottoman fleet in Suez.[2] Today, the Topkapı Palace Library holds only the western part of his world map, which depicts the Atlantic Ocean, the western coast of Europe and Africa, and the New World. It has long been argued that Selim used the missing part of the map—which shows the Mediterranean and the Indian Ocean—in the planning of his future conquests.[3]

Selim I did not appoint Piri Reis as "official cartographer." This, however, did not stop Piri Reis or his colleagues interested in the geography of the world. They charted the seas, islands, and port cities of the Mediterranean. They narrated the distant parts of the world and their inhabitants. They described the lands of the Ottoman Empire. These Ottoman intellectuals combined the Islamic and European traditions of geographical knowledge and formulated the Ottoman canonical geographical knowledge. In preparing these accounts, they satisfied their own curiosity, displayed their skills in navigation, cartography, and writing, and at the same time sought to earn the favors of the Ottoman sultan or a member of his court. As the number of geographical accounts circulating in Constantinople increased, the Ottoman court started frequently extending its patronage to these intellectuals.

This book is the story of these geographers and the rich body of geographical literature they left behind for future generations. In the absence of an institution regulated and financed by the Ottoman state, these individuals endeavored to consolidate the geographical knowledge in the Ottoman Empire on the one hand and to earn the favor of the Ottoman court on the other. For the most part, they were patronized by the members of the Ottoman court who tried to shape imperial policy. These geographers became the "eyes of the world" for the Ottoman Empire. They projected the world in a fashion that was easily accessible

---

2 Svat Soucek, "Piri Reis," *EI*[2], last accessed via Brill Online. University of Pittsburgh. 7 November 2011.

3 Adnan Adıvar, *Osmanlı Türklerinde İlim*, 5th ed. (Istanbul: Remzi Kitabevi, 1991), 76.

to the court. In their works, they synthesized the traditional accounts of geography with their own observations. They employed traditional styles such as travel accounts or coordinate tables to represent new knowledge. Nevertheless, they also did not refrain from employing novel means such as *portolan* charts and *isolarii* (island books). These individuals' endeavors generated the Ottoman style in the field of geography. The sixteenth century witnessed the most dramatic expansion of the Ottoman Empire in its history. In the 1600s, it was the largest empire in the Old World and was broadcasting its imperial claims to universal sovereignty in both the East and the West. Ottoman geographers helped formulate this enterprise by providing the members of the Ottoman court with practical geographical information. The Ottoman court was aware of the symbolic and practical value of the geographical works and took advantage of them.

In the sixteenth century, Europe saw a surge in the number of travel accounts and maps depicting the Mediterranean, the Far East, the Ottoman domains, and the New World. Sparked by the "Age of Exploration," these accounts were among innumerable geographical works that redefined for Europeans the boundaries of the inhabited world in this period.[4] In the case of the Spanish Habsburgs and the Portuguese, European courts also established institutions to regulate the production and distribution of geographical works. Although it is well known that rulers in medieval and early modern Europe commissioned geographical works to project their royal or imperial aspirations,[5] the rich collection of surviving Ottoman geographical accounts has never received the same analytical treatment, presumably because the Ottomans did not have a centralized institution for geography and did not participate in the so-called "Age of Exploration." While recent studies demonstrate that the Ottomans not only participated in but also helped to define a variety of early modern diplomatic, economic, and religious-political trends,[6]

[4] Sanjay Subrahmanyam, "Connected Histories: Notes towards a Reconfiguration of Early Modern Eurasia," *Modern Asian Studies* 31 (1997): 737.

[5] J. B. Harley, "Silences and Secrecy: The Hidden Agenda of Cartography in Early Modern Europe," *Imago Mundi* 40 (1988): 57–76; David Buisseret, ed., *Monarchs, Ministers, and Maps: The Emergence of Cartography as a Tool of Government in Early Modern Europe* (Chicago: University of Chicago Press, 1992).

[6] For Ottoman activities in the Indian Ocean and their effects on world politics in the early modern period, see Giancarlo Casale, *The Ottoman Age of Exploration* (Oxford: Oxford University Press, 2010). For the Ottomans' place in the early modern world in general, see Subrahmanyam, "Connected Histories," 735–62; Lisa Jardine and Jerry Brotton, *Global Interests: Renaissance Art between East and West* (Ithaca: Cornell University Press, 2000); Daniel Goffman, *The Ottoman Empire and Early Modern World* (New York: Cambridge University Press, 2002); Gerald MacLean, *Re-Orienting the Renaissance: Cultural Exchanges with the East* (New York: Palgrave Macmillan, 2005); Matthew Birchwood and Matthew Dimmock, eds., *Cultural Encounters between East and West, 1453–1699* (Cambridge: Cambridge Scholars Press, 2005); Gábor Ágoston, "Information, Ideology, and Limits of Imperial Policy: Ottoman Grand Strategy in the Context of Ottoman-Habsburg Rivalry," in *Early Modern Ottomans: Remapping the Empire*, ed. Virginia Aksan and Daniel Goffman (New York: Cambridge University Press, 2007), 75–103.

Ottoman geographical literature has been analyzed only on rare occasions by historians for its value as a tool of imperial propaganda.[7]

This book maintains that the Ottoman Empire was an active participant of the Early Modern Period and of the "Age of the Exploration." In the sixteenth century, the empire's presence in the Eastern Mediterranean, Southeast Europe, and on the shores of the Indian Ocean made its court and intellectuals indispensable members of the global political, economic, and intellectual networks. Ottoman geographers were among the group of intellectuals who most actively participated in these networks. In their works, they combined the traditional Muslim accounts of geography with their first-hand observations and recent European geographical accounts that circulated in the Mediterranean. They had two aims in preparing these accounts for the Ottoman court. First, they wanted to display their talent and knowledge to the court and thus in return earn favors and patronage from its members. Second, they wanted to provide practical and symbolic support for the developing Ottoman imperial claims.This was a valient effort on the part of Ottoman geographers. Although the Ottoman court had an official historian since the fifteenth century, it never considered appointing an official geographer. However, the tenacious efforts of these geographers to inform the Ottoman court about the latest geographical knowledge did not go unrecognized. While many of them gained favors from the members of the imperial court in Constantinople, their works helped shape the Ottoman imperial policies.

[7] For a general treatment of the Ottoman cartographic tradition, see Ahmet Karamustafa, "Introduction to Ottoman Cartography," in *The History of Cartography, Vol. 2, Bk. 1, Cartography in the Traditional Islamic and South Asian Societies*, ed. J. B. Harley and David Woodward (Chicago: University of Chicago Press, 1992), 205–8; Svat Soucek, "Islamic Charting in the Mediterranean," in ibid., 263–87. On Ottoman travel literature, see Nicolas Vatin, "Pourquoi un Turc Ottoman Racontait-il son Voyage? Note sur Les Ottomans des Vaki'at-i Sultan Cem au Seyahatname d'Evliya Çelebi," in *Les Ottomans et L'Occident (XVe – XVIe siècles)* (Istanbul: Les Éditions ISIS, 2001), 179–93. On the relationship between Ottoman imperial politics and geographical works, see Jerry Brotton, *Trading Territories: Mapping the Early Modern World* (Ithaca: Cornell University Press, 1997); Kathryn Ebel, "City Views, Imperial Visions: Cartography and the Visual Culture of Urban Space in the Ottoman Empire, 1453–1603" (PhD diss., University of Texas at Austin, 2002); Karen C. Pinto, "Ways of Seeing: Scenarios of the World in the Medieval Islamic Cartographic Imagination" (PhD diss., Columbia University, 2001); Pınar Emiralioğlu, "Cognizance of the Ottoman World: Visual and Textual Representations in the Sixteenth-Century Ottoman Empire (1514–1596)" (PhD diss., University of Chicago, 2006); Casale, *The Ottoman Age of Exploration*. For a comprehensive analysis of Ottoman geographical studies and influences of Arab geographical traditions on Ottoman intellectuals, see Gottfried Hagen, *Ein osmanischer Geograph bei der Arbeit Entstehung und Gedankenwelt von Katib Celebis Cihannüma* (Berlin: Klaus Schwarz Verlag, 2003), 79–119.

Ottoman Geographical Consciousness and Sixteenth-Century Imperial Enterprise

This study demonstrates that there was a heightened sensitivity to geographical knowledge, or in other words, scholarly and intellectual engagement in understanding and knowing the world's geography, in the sixteenth-century Ottoman Empire.[8] This development was intimately related to the consolidation of Ottoman claims to world conquest and universal imperial sovereignty. The Ottoman Empire shared political, cultural, economic, and religious conceptual frameworks—such as the notions of universal monarchy, justice, cultural and religious unity through imperial patronage, and territorial expansion through "just wars"—with contemporary and previous world empires.[9] Although canonical literature ties the beginnings of Ottoman imperial ideology to Mehmed II's conquest of Constantinople in 1453, recent studies show that a distinct imperial Ottoman enterprise began to develop only in the sixteenth century.[10] This new imperial vision underscored the Ottoman desire to become a world empire, and positioned the empire in relation to its rivals in both East and West.

Among the intellectuals involved in this shift toward imperialism was a group of travelers, historians, sea captains, cartographers, and astrologers who together created a distinct body of geographical literature in the sixteenth century. Most of these geographers were products of the Ottoman administration. They were trained in either the palace schools or madrasas around the Muslim world. These bureaucrats and members of the Ottoman military establishment worked under or sought the patronage of the Ottoman court, which included the sultan, his household, and members of his imperial council. These geographers frequently traveled within and beyond the Ottoman realm and were aware of each other's experiences. They were also conversant with the Muslim geographical traditions and the latest geographical literature—in particular, maps—circulating in Europe

---

[8] For a discussion on the separation between "geography as a subject and geographical knowledge as a set of intellectual practices concerned with knowing the world" within the context of the eighteenth century, see Charles Withers and Robert J. Mayhew, "Geography: Space, Place and Intellectual History in the Eighteenth Century," *Journal for Eighteenth Century Studies* 34, no. 4 (2011): 445–6.

[9] See especially Anthony Pagden, *Lords of All the World: Ideologies of Empire in Spain, Britain and France c. 1500–c. 1800* (New Haven: Yale University Press, 1995); Barbara Bush, *Imperialism and Postcolonialism* (London: Pearson Education, 2005); Sanjay Subrahmanyam, "Connected Histories."

[10] Gülru Necipoğlu, "Süleyman the Magnificent and the Representatiom of Power in a Context of Ottoman-Habsburg Papal Rivalry," *Art Bulletin* 71 (1989): 401–27; idem, "A Kanun for the State, a Canon for the Arts," in *Soliman le Magnifique et son Temps*, ed. Giles Veinstein (Paris: La Documentation Française, 1992), 195–216; Cornell Fleischer, "The Lawgiver as Messiah: The Making of the Imperial Image in the Reign of Süleyman," in *Soliman le Magnifique et son Temps*, ed. Giles Veinstein (Paris: La Documentation Française, 1992), 159–77.

and in the Mediterranean. Using this knowledge as a foundation, they created a large body of geographical works—maps, travel accounts, charts, and geographical treatises—of the known world in the sixteenth century.

In addition to the geographical works prepared by Ottoman geographers, the Ottoman court also commissioned maps from European cartographers. Sixteenth-century Constantinople was not oblivious to the political and intellectual developments taking place in Europe. In addition to the reports they received from the influential Jewish subjects who usually served as confidants to the Ottoman court, the sultan and the grand vizier gathered intelligence from the European envoys sent to Constantinople and from the staff of the European embassies in the city.[11]

The scope and number of Ottoman geographical works created in the sixteenth century never reached the level of those produced for the Portuguese and Spanish courts. Nor did the Ottoman court ever institutionalize the production of geographical works. Nevertheless, it welcomed and encouraged individual geographers to develop a canonical body of geographical literature that was accessible and comprehensible. These works were almost exclusively penned in Ottoman Turkish. This was a new chapter in the history of the geographical works in the Muslim world. The traditional historiography on Islamic geography tends to dismiss the developments in the Ottoman Empire and marks the fifteenth century as the end of original geographical works in the Muslim world. It is the contention of this book that the production and diffusion of geographical works reached unprecedented levels in the sixteenth century. Ottoman geographers actively participated in this endeavor. They translated Arabic and Persian works into Ottoman Turkish and also produced original works of geography in a variety of forms and on a variety of topics. In the process, they found opportunities for patronage among the members of the Ottoman court. In return, these works enhanced the court's geographical understanding of the world and gave impetus to its imperial campaigns, especially in the Mediterranean and in Southeast Europe.

Even more importantly, these geographical works could be used to buttress the political agendas of their patrons. Those prepared for the Ottoman sultan narrated and depicted the geographical features of the Ottoman realm and of different parts of the world. In doing so, they sought to portray the Ottoman sultans as universal sovereigns. They promoted the imperial magnificence of the Ottoman sultans, and projected the Ottoman Empire as the center of the universe. It should be noted that the Ottoman geographers' portrayals of their empire as the hub of the universe were not without precedent. Throughout the ages, Chinese, European, Indian, and Islamic civilizations attempted to place themselves at the center of history and maps.[12] The Ottoman Empire was no exception. Nevertheless, Ottoman geographers challenged medieval and early modern Muslim and Christian traditions. These geographers, like their counterparts in other

[11] Ágoston, "Information, Ideology, and Limits of Imperial Power," 82–3.

[12] Marshall G. S. Hodgson, *Rethinking World History: Essays on Europe, Islam and World History*, ed. Edmund Burke (Cambridge: Cambridge University Press, 1993).

contemporary empires, systematically engaged in the geography of the world and produced a variety of geographical accounts that illuminated their central position at the intellectual and political networks of the time.

Defining the Boundaries of the Ottoman World

The geographical works analyzed in this study were prepared and distributed in the Ottoman Empire during the reigns of Selim I (r. 1512–20), Süleyman I (r. 1520–66), and Süleyman's successors Selim II (r. 1566–74) and Murad III (r. 1574–95). These were decades of dramatic territorial expansion, especially during the reigns of Selim I and Süleyman I. Knowing the location and characteristics of newly conquered territories and lands beyond the Ottoman domains was essential. For this reason, maps and travel accounts had significant geopolitical implications. Ottoman geographers were crucial actors in the territorial expansion of the empire's domains during this period. They created a separate body of geographical literature and consciously integrated both the newly conquered areas and the frontiers of the known world into their and their rulers' geographical knowledge. Thus, Ottoman geographers supported imperial expansion through the practical information they provided. At the same time, they broadcast Ottoman imperial claims to their political and religious rivals in the East and West.

This book offers the historical analysis of all the available geographical accounts that circulated in Constantinople in the sixteenth century. These include *portolan* charts and atlases, world maps, travel accounts, and coordinate tables. It should be noted here that Ottoman geographical accounts from this period are not as numerous as European accounts. Therefore, in order to chart a comprehensive map of social and political history of geographical knowledge in sixteenth-century Ottoman Empire, this book also makes use of Ottoman chronicles and campaign diaries all of which offer insights about the Ottoman understanding of the world around them.

For interpretive background, this book draws upon the field of historical geography, particularly the literature that emphasizes the role of geographical works as tools of propaganda in the early modern world.[13] The "hidden agenda" of cartographical and geographical works has been at the center of studies on historical geography since the late 1980s. The pioneer in this field, J. B. Harley, highlighted the importance of "maps as an instrument of power" in the context of early modern Europe and their strong connection to the dominant ideology of

---

[13] For some works that discuss the relationship between history and geography and the relationship between state ideology and geographical works, see J. B. Harley, "Silences and Secrecy"; David Buisseret, ed., *Monarchs, Ministers, and Maps: The Emergence of Cartography as a Tool of Government in Early Modern Europe* (Chicago: University of Chicago Press, 1992); J. B. Harley and David Woodward, ed., *The History of Cartography, Vol. 2, Bk. 1, Cartography in the Traditional Islamic and South Asian Societies* (Chicago: University of Chicago Press, 1992); Alan R. H. Baker, *Geography and History: Bridging the Divide* (Cambridge: Cambridge University Press, 2003), 1–36.

central authorities.[14] Here, Harley applied Foucault's analysis on the relationship between power and knowledge to cartography. During the last decades, research by a number of historians who have refined Harley's methodology has strengthened the understanding of geographical works as tools for central powers to project their world perspectives. These studies also introduced new avenues to examine the ideological and cultural dimensions of map production and its links to the economic and social policies in the early modern period.[15] As this book demonstrates, by reconstructing the production and distribution of the large corpus of Ottoman geographical accounts, it is possible to situate the practice of geography, and more specifically the role of Ottoman geographers, within the broad comparative context of the early modern world.

Ottoman geography has been underrepresented in this field. From the second volume of the *History of Cartography* we know that Ottoman geographers developed a distinct Ottoman tradition in their creation of charts, military plans, town views, and travel accounts starting as early as the fifteenth century.[16] From the fifteenth to the mid-seventeenth century, this tradition was mainly based on classical Arabic and Persian sources. According to the literature, Ottoman geographers then came under the influence of Europeans from the mid-seventeenth century onwards.[17] Recent studies have called attention to the importance of Ottoman geographical works as tools of imperial propaganda.[18] This study adds to that literature and argues that Ottoman geographical and imperial expansion were deeply entwined in the sixteenth century.

The organizing scheme of the book is the framework by which the Ottoman geographers categorized the known world for their patrons. In their accounts, they portrayed Constantinople as the new nucleus from which power radiated

14 J. B. Harley, "Silences and Secrecy."

15 Barbara Mundy, *The Mapping of New Spain: Indigenous Cartography and the Maps of the Relaciones Geograficas* (Chicago: University of Chicago Press, 1996); Matthew Edney, "The Irony of Imperial Mapping," in *The Imperial Map: Cartography and the Mastery of Empire*, ed. James Akerman (Chicago: University of Chicago Press, 2009), 11–45; Valerie Kivelson, *Cartographies of Tsardom: The Land and Its Meanings in Seventeenth-Century Russia* (Ithaca: Cornell University Press, 2006); Neil Safier, "The Confines of the Colony: Boundaries, Ethnographic Landscapes, and Imperial Cartography in Iberoamerica," in *The Imperial Map*, ed. James Akerman (Chicago: University of Chicago Press, 2009), 133–85.

16 Ahmet T. Karamustafa, "Military, Administrative, and Scholarly Maps and Plans," in *The History of Cartography, Vol. 2, Bk. 1, Cartography in the Traditional Islamic and South Asian Societies*, ed. J. B. Harley and David Woodward (Chicago: University of Chicago Press, 1992), 209–28; Svat Soucek, "Islamic Charting in the Mediterranean," in ibid., 263–87; J. M. Rogers, "Itineraries and Town Views in Ottoman Histories," in ibid., 228–56.

17 Karamustafa, "Military, Administrative, and Scholarly Maps and Plans," 218; Rogers, "Itineraries and Town Views in Ottoman Histories," 228–9.

18 Ebel, "City Views, Imperial Visions"; Pinto, "Ways of Seeing"; Casale, *The Ottoman Age of Exploration*; Brotton, *Trading Territories*, 87–119.

throughout the Ottoman domains and beyond. They divided the rest of the world into "core" and "periphery." Here, I adopt terms from dependency theory and its application in the world-system analysis laid out by Immanuel Wallerstein in 1974.[19] The world-system in Wallerstein's analysis is characterized by a hierarchy of regions where cores and peripheries interacted. These interactions produced economic inequalities between regions and within the interconnected societies.[20] The Ottoman court and geographers viewed the world in hierarchical terms as well, but their analytical perspectives were quite different. In their scheme, the core areas were the regions where the Ottoman sultans had to substantiate their imperial claims. The peripheral areas—India, the Indian Ocean, and the New World—were the regions geographically located on the frontiers. Although geographical as well as social and political information about these places had to be assimilated into already existing knowledge to protect and preserve the empire, in the Ottoman court's political vision these were not areas where Ottoman imperial claims had to be proven. It was within this context that Ottoman geographers presented the Mediterranean as the core of the Ottoman world.

The point of departure for this book is a series of questions that have not been posed in other early modern studies. What historical circumstances triggered the Ottoman claims to universal sovereignty? How did they influence the production and distribution of geographical works in the Ottoman Empire? What was the relationship between the rise of Ottoman canonical geographical discourse and imperial and political discourse? What does this relationship in return tell us about the early modern Ottoman world?

This book begins with a study of political and cultural interactions between the sixteenth-century Ottoman Empire and the Habsburg Empire in the West, and the Ottoman Empire and the Safavid Empire in the East. It goes on to examine the scholarly and intellectual practice of sixteenth-century Ottoman geography. This involves a survey of the discipline's principal texts, particularly Ottoman chronicles (*Tevarih-i Al-i Osman*), royal histories (*shahname*), campaign diaries (*menazilname*), and travel accounts, as well as cartographical works produced and distributed within the Ottoman Empire. The result is an examination of how the Ottoman Empire dealt mentally and politically with territorial expansion in the sixteenth century, and how it assimilated this conquered space into the already known territories.[21] Although the following chapters will present detailed textual analysis of the geographical works and delineate thoroughly how they shaped

---

19 Immanuel Wallerstein, *The Modern World System I: Capitalist Agriculture and the Origins of the European World-Economy in the Sixteenth Century* (New York and London: Academic Press, 1974).

20 Wallerstein, *The Modern World System I*, 347–50.

21 For a discussion on the early modern Ottoman space and its construction in early modern European and Ottoman sources, see Palmira Brummett, "Imagining the Early Modern Ottoman Space, from World History to Piri Reis," in *The Early Modern Ottomans: Remapping the Empire*, ed. Virginia H. Aksan and Daniel Goffman (Cambridge: Cambridge University Press), 15–58.

the Ottoman court's political agenda and emphasized the importance of the Mediterranean for the imperial enterprise, the aim in the first chapter is to offer a glimpse of the development of geographical literature in the Ottoman Empire in the sixteenth century. This chapter also sets up a dialogue among Ottoman and Spanish Habsburg geographical works, and shows how the empires in the early modern world territorialized their imperial claims.

The chapters that follow are organized thematically and move from Constantinople, the nucleus, to the Mediterranean, the core, and finally to the Indian Ocean and the New World, the peripheries of the Ottoman Empire. The second chapter situates Ottoman geographical consciousness in the broader context of the development of Ottoman imperial rivalry with the Habsburgs and Safavids during the first half of Süleyman the Magnificent's reign (r. 1520–66). As part of the new imperial enterprise, the Ottoman court redesigned Constantinople and highlighted it as the new imperial center. This chapter investigates how Ottoman geographers from the reign of Mehmed II on participated in this process. It examines such key works as Mustafa b. Ali al-Muvakkit's (d. 1571) *İ'lamü'l-'İbad fi A'lami'l-Bilad* (*Public Instructions for the Distances of Countries*), Abdüllatif's (Latifi) (d. 1582) *Evsaf-ı İstanbul* (*Qualities of Istanbul*), and Matrakçı Nasuh's (d. 1564) *Mecmu'-i Menazil: Beyan-ı Sefer-i Iraqeyn-i Sultan Süleyman Han* (*The Collected Stages: The Description of Sultan Süleyman's Campaign in the Two Iraqs*) from 1537. Analysis of these works reveals that the Ottoman geographers were genuinely interested in updating knowledge of Anatolia, the Balkans, and the Near East. They recalculated the coordinates of important cities situated in these three areas, and provided detailed topographical and literary accounts. In doing so, they reoriented the world around Constantinople.

The third chapter of the book focuses on the reign of Süleyman the Magnificent and on the ways in which Ottoman policymakers sought to counterbalance Charles V and Philip II's claims to universal Christian rulership, as well as the Safavid Shah Tahmasb's assertions regarding sovereignty over Shi'a Ottoman subjects, by employing two key aspects of rulership: imperial law and control of the Mediterranean. In the early modern period, the prevailing model of imperial power was still Roman. Without control of the Mediterranean, universal imperial power could not be imagined or convincingly projected. In the sixteenth century, the Ottoman Empire actively participated in the larger imperial trends of the time and thereby helped to define the Spanish Habsburg's diplomatic, economic, and religious policies. Opposing empires at each end of the Mediterranean shared common imperial, political, and intellectual concepts of power. In general, this chapter argues that for Ottoman geographers and imperial policymakers, the Mediterranean was the core of the world. Here, I present the Ottoman geographer and sea captain Piri Reis's (d. 1554) navigational guidebook, the *Kitab-ı Bahriye* (*Book of Sea Lore*) of 1526, as an example of "official" Ottoman cartography.

The book's fourth chapter looks at the relationship between two peripheries: the geographical boundaries of the known world and the peripheries of Ottoman imperial power. By focusing on visual and textual depictions of India, the Indian

Ocean, and the New World, this chapter investigates how Ottoman geographers marked the empire's boundaries and established connections to distant regions of the world. Here, Piri Reis's world map (1513), presented to Selim I in 1517, Seydi Ali Reis's *Mir'atü'l-Memalik* (*Mirror of Lands*) and the anonymous account on the New World, *Tarih-i Hind-i Garbi* (*History of the India of the West*), are significant examples of the Ottoman court's openness to the latest geographical and political knowledge about faraway lands and Ottoman geographers' efforts to bring these peripheries closer to the center of imperial power through intellectual means. This chapter maintains that the Ottoman intellectual and political interest in the Indian Ocean was much less systematized and organized than has so far been argued. At the same time, the New World was a region in which the Ottoman court showed tangential, yet increasing awareness.

Finally, the epilogue carries the history of Ottoman geographical knowledge into the long eighteenth century. By focusing on a select number of geographical and cartographical accounts between the 1650s and 1802, the final chapter of this book offers preliminary insights into how Ottoman geographical knowledge changed from the sixteenth to the eighteenth century. Here, Katip Çelebi's (d. 1657) work on world geography, *Cihannüma*, and Ebu Bekir ibn Behram ed-Dimashki's (d. 1691) translation of Dutch cartographer Joan Bleu's *Atlas Maior* are analyzed as examples of Ottoman geographical works that aimed to educate the Ottoman court as well as the larger audience of literate urban dwellers. This chapter proposes that eighteenth-century Ottoman geographical works paint a vivid picture of active and conscious Ottoman involvement in contemporary intellectual and political discussions. It demonstrates that in the eighteenth century, informal and borderless endeavors of individual geographers of the sixteenth century yielded to formal and more conscious efforts by Ottoman geographers to satisfy the requirements of the science of geography as well as the changing political and economic realites of the period.

Chapter 1
Negotiating Space and Imperial Ideology in the Sixteenth-Century Ottoman Empire

The first half of the sixteenth century was a period of rapid territorial expansion that put the Ottomans in contact with different geographical regions and peoples. New geographies and the urgent need to consolidate their control over newly conquered lands and peoples forced a change in the Ottomans' understanding of space and empire. Triggered by military conflict and territorial expansion in the East and West, these changes revolutionized the political-cultural perspectives of the Ottoman ruling elites and intellectuals. This new era in the history of the Ottoman Empire coincided with the reigns of Selim I (r. 1512–20) and Süleyman I (r. 1520–66). During the reign of these sultans, the Ottomans reached Hungary and took control of strategic islands and ports in the Mediterranean. In the East, the Ottoman domains extended to the Red Sea, with the potential to reach the shores of the Indian Ocean. Assimilating the newly conquered territories and peoples into the Ottoman domains was simultaneously a political and mental process. As the Ottoman court articulated its sultan's role as a world conqueror and the sovereign of an ever-expanding empire, intellectuals revised the existing geographical literature to either suit or shape an emerging vision of the empire.

This Ottoman imperial project did not take place in a vacuum but rather in a context shaped by interaction with the Habsburg Empire in the West and the Safavid Empire in the East. The Ottomans shared a variety of imperial cultural and political strategies with these rival empires, with whom they engaged militarily, politically, and intellectually throughout the sixteenth century. This chapter analyzes these interactions and demonstrates how military and political conflicts with the Habsburgs and the Safavids influenced Ottoman geographical consciousness. It offers a close textual analysis of Ottoman chronicles, royal histories, campaign diaries, travel accounts, and geographical works. One of this chapter's central aims is to show that a distinct group of geographers emerged in the Ottoman Empire during the sixteenth century. This select group of intellectuals informed their elite audience, the members of the Ottoman court, on the latest geographical knowledge. In doing so, they tried to shape the political visions of their audience and promote the agenda of their patrons. Understanding the processes of production, the purpose, use, and reception of these geographical accounts will shed light on the intellectual and political networks interlinked in the sixteenth century Ottoman Empire. In this period, Ottoman geographers published works that displayed their knowledge and talents and sought patronage from the members of the Ottoman court. The Ottoman court, in return, commissioned maps, travel accounts, and cosmographies to articulate their imperial claims to the Ottoman center and beyond.

Selim I and the Formation of Ottoman Imperial Ideology

> It shall be known that the superior aim and highest claim of the establishment of the divine caliphate is restricted to empowering the Muslim religion and fending off the vestiges of oppression and erasing unbelief and heresy from the lands of Cairo [...] and by the existence of its glory and pride to serve the holy places and to protect the principles of Islam and provide for purity and unity.[1]

Two years after ascending to the Ottoman throne in 1514, Sultan Selim I embarked upon one of the most difficult military campaigns in Ottoman history. His rival was Shah Ismail of the Safavids, who was then in nominal control of Tabriz, Ardabil, Yazd, Kashan, Hamadan, Isfahan, Shiraz, Diyar Bakr, and Baghdad. Ismail, the young and charismatic leader of the Safavid Sufi order, had captured Tabriz in 1501 and declared the city his capital.[2] His rapid territorial expansion posed a threat to the Ottoman domains. To make matters worse, Ismail proclaimed Twelver Shi'ism as his new realm's religion, which placed even more pressure on the Ottomans by introducing a religious dimension into the conflict. Ismail's ascendancy had already affected the Ottoman domains during the reign of Selim's father, Bayezid II (r. 1481–1512). At the beginning of the sixteenth century, Safavid missionaries tried to influence the Turkmen tribes in eastern Anatolia. Since the reign of Mehmed II, the Ottoman state had been trying to bring the tribal population in eastern Anatolia under the central administration and break up the old order of tribal allegiances.[3] Unhappy with the Ottoman state's taxation and land administration policies, the Turkmen tribes welcomed Ismail's patronage and Safavid appreciation for their nomadic way of life and readiness to engage in warfare. Selim, who was then a prince in Trebizond, had witnessed these growing

---

1 Selim I's letter to the Shirvanshah, the rulers of the Shirvan region in present-day Azerbaijan, after the Ottoman victory against the Mamluks in 1516 and 1517. Feridun Ahmed Beg, *Münşeatü's-Selatin* (Istanbul, 1858), vol. 1, 438–9. Translations are mine unless otherwise stated.

2 For the role of Ismail in the creation of the Safavid state and the various personalities he adopted, see Jean Aubin, "Études Safavides. I. Shah Isma'il et les notables de l'Iraq persan," *JESHO* 2 (1959): 37–81; idem, "La Politique religieuse des Safavids," in *Le Shiisme imamite*, ed. T. Fahd (Paris: Presses Univ. de France, 1970), 235–44; idem, "L'Avenèment des Safavides Reconsidéré (Études Safavides. III)," *Moyen Orient et Ocean Indien* 5 (1988): 1–130; V. Minorsky, "The Poetry of Shah İsma'il I," *BSOAS* 10 (1940–42): 1006a–1053a; Colin P. Mitchel, *The Practice of Politics: Power, Religion, and Rhetoric* (London and New York: I. B. Tauris, 2009), 19–58; Andrew Newman, *Safavid Iran: Rebirth of a Persian Empire* (London: I. B. Tauris, 2006), 13–25.

3 Jean-Louis Bacqué-Grammont, *Les Ottomans, les Safavides, et leurs voisins* (Istanbul: Nederlands Historisch-Archaeologisch Instituut te Istanbul, 1986), 17–18. On the influence of Selim's land policies on the tribes of Anatolia also see Irène Beldiceanu-Steinherr and Jean-Louis Bacqué-Grammont, "A Propos de quelques causes de malaises sociaux en Anatolie Centrale," *Archivum Ottomanicum* 7 (1982): 71–116.

religious and military challenges to Ottoman legitimacy, and he vowed to repel anyone who threatened the Ottomans or encroached on their territory.[4]

In the summer of 1514, Selim's armies defeated Shah Ismail at the Battle of Chaldıran. No sooner had he entered the Safavid capital of Tabriz in September 1514 than he immediately launched his next campaign against the Mamluks of Egypt and Syria.[5] After capturing Aleppo and Damascus in 1516, Selim I ordered immediate preparations for the invasion of Egypt—he entered Cairo in 1517. Following this final triumph, the *sharif*s (descendants of the Prophet Muhammad and governors) of Mecca and Medina recognized Selim's sovereignty.[6]

At first glance, these campaigns, which resulted in the addition of new territories to the Ottoman domains, appeared to be perfect opportunities for Selim to trumpet his talents as a soldier and ruler. But victory presented bigger challenges than Selim had anticipated. During and after these campaigns, Selim I was confronted with a difficult task. While campaigns against Christian empires were noncontroversial given Christendom's demonstrable hostility to Islam and followed a well-established tradition of frontier warfare known as *ghaza*, campaigns against Muslim powers presented unique challenges.[7] When the Ottomans expanded into Anatolia in the fourteenth and early fifteenth centuries, they fought against Muslim principalities such as Karasi. But, the early Ottomans were still early converts. They moved and acted within the liminal spaces between the monotheistic faiths. In this period, there was no overarching political or religious structure to define how the state religion should be observed or which actions by the rulers should be considered against Islam.[8] However, the political conditions had now changed. War with the Safavids in the sixteenth century clashed with the Ottomans' social, economic, and ideological expectations. Now, Selim would have to legitimize his actions before the Janissaries, Ottoman subjects, and neighboring Muslim rulers. Selim simply could not ignore the Safavid threat. Safavid military power and Shi'i proselytizing activities put Ottoman sovereignty in eastern Anatolia at risk. Shah Ismail's charisma and popularity in the region prompted the Ottoman ruler to obtain a *fatwa* (legal opinion) from his

---

4 On Selim's rivalry against his brothers, see Çağatay Uluçay, "Yavuz Sultan Selim Nasıl Padişah Oldu?" *Tarih Dergisi* 9 (1953): 53–90; 10 (1954): 117–42; 11 (1955): 187–200; Şahabettin Tekindağ, "Şah Kulu Baba Tekeli İsyanı," *Belgelerle Türk Tarihi Dergisi* 3 (1967/68): 34–9; 4 (1967/68): 54–9; Ahmed Uğur, *The Reign of Selim I in the Light of the Selim-name Literature* (Berlin: K. Schwarz, 1985); Hakkı Erdem Çıpa, "The Centrality of the Periphery: The Rise to Power of Selim I, 1487–1512" (PhD diss., Harvard University, 2007).

5 Selahattin Tansel, *Yavuz Sultan Selim* (Ankara: Milli Eğitim Basımevi, 1969), 73–81; 101–7.

6 Halil İnalcık, "Selim I," *EI²*, last accessed via Brill Online. University of Pittsburgh. 7 November 2011.

7 On *ghaza* ideology, see Colin Imber, "Ideals and Legitimation in Early Ottoman History," in *Suleyman the Magnificent and His Age: The Ottoman Empire in the Early Modern World*, ed. Metin Kunt and Christine Woodhead (London: Longman, 1995), 138–53.

8 Cemal Kafadar, *Between Two Worlds: The Construction of the Ottoman State* (Berkeley and Los Angeles: University of California Press, 1996), 74–7. Also see Daniel Goffman, *The Ottoman Empire and Early Modern Europe* (Cambridge: Cambridge University Press, 2002), 41–2.

Şeyhülislam (the grand mufti of Constantinople), declaring Ismail and his followers to be heretics whose destruction was not merely legitimate but obligatory.[9]

Shah Ismail openly challenged the Ottoman sultan's worldly and spiritual authority. His sympathizers proliferated in Anatolia and tried to subvert Selim's political authority in the region. Ismail's own poetry reveals that the young shah used Persian and Mediterranean cultural references and Twelver Shi'ism simultaneously to create an image of and for himself. In his poems, he referred to himself as Faridun, Khusraw, Rustam, Alexander, and Jesus and he also claimed to be the *Mahdi* (the guided one), who would emerge from hiding and drastically reform the world.[10] By openly claiming to be the *Mahdi*, Ismail combined political and religious authority in himself.[11]

Shah Ismail's image as challenger to the mighty Selim I resonated in Europe, traveling beyond the Ottoman and Safavid domains. In Venice and France, reports about the Safavid Shah emphasized his sympathetic stance toward Christians. In these reports, Ismail was deemed a messianic figure who would stop Ottoman expansion from the east.[12] European powers thus sought economic and political alliances with the Safavids, grateful for any potential check to the Ottomans' seemingly unstoppable western expansion. Selim had reasons to believe that enemies surrounded the Ottoman realm, a situation that military might alone could not resolve.

The Ottoman response to the Safavid threat generated a new political and cultural discourse in Constantinople. After conquering Egypt and Syria, Selim became an active champion of Islam, defending his realm against the "heretics" and "infidels"—in other words, Safavids and Mamluks—whose intransigence hindered the empire's struggle against its Christian enemies.[13] In response to Shah

[9] Colin Imber, *The Ottoman Empire, 1300–1650* (New York: Palgrave Macmillan, 2002), 45.

[10] *Mahdi* is an apocalyptic term meaning "the guided one" who will descend before the End of Time and kill *Deccal*, one of the last signs of that time's arrival.

[11] Said Amir Arjomand, *The Shadow of God and the Hidden Imam: Religion, Political Order and Societal Change in Shiite Iran from the Beginning to 1890* (Chicago: University of Chicago Press, 1984), 109. For a discussion on the similarities between the Ottoman and Safavid discourses in this period, see Markus Dressler, "Inventing Orthodoxy: Competing Claims for Authority and Legitimacy in the Ottoman-Safavid Conflict," in *Legitimizing the Order: The Ottoman Rhetoric of State Power*, ed. Hakan T. Karateke (Leiden: Brill, 2005), 151–73.

[12] Palmira Brummett, "The Myth of Shah Ismail Safavi: Political Rhetoric and 'Divine' Kingship," in *Medieval Christian Perceptions of Islam*, ed. John Victor Tolan (New York: Garland, 1996), 336; on the image of Shah Ismail in sixteenth-century Venetian humanist writings, see Margaret Meserve, *Empires of Islam in Renaissance Historical Thought* (Cambridge, MA: Harvard University Press, 2008), 231–7.

[13] Gábor Ágoston, "Information, Ideology, and Limits of Imperial Policy: Ottoman Grand Strategy in the Context of Ottoman-Habsburg Rivalry," in *Early Modern Ottomans: Remapping the Empire*, ed. Virginia Aksan and Daniel Goffman, 93. This can best be seen in Selim's letters to the rulers of Central Asia and Crimea following the 1514 victory. Feridun Ahmed Beg, *Münşeatü's-Selatin*, vol. 1, 386–95.

Ismail's embodiment of important Mediterranean religious and political figures, the Ottoman sultan and his ruling elites began to refer to Selim as "the protector of the two holy cities and the pilgrimage routes, and of all Muslims of the world," the *Mahdi*,[14] and *sahib-kıran-ı heft iklim* (master of the auspicious conjunction in seven climes), pertaining to his status as universal ruler.[15] These claims by Selim I not only challenged Ismail's image and popularity but also quieted the unruly troops. Thus began the transformation of the office of the sultan from a temporal political ruler to an idealized spiritual force. Some contemporary historians believed that if Selim had lived longer, he would have become the next Alexander the Great or Genghis Khan.[16] The Ottoman Empire became a major player in the Mediterranean and the Red Sea through his conquests.

Selim's World: The Mediterranean and the Red Sea

Selim's conquests extended Ottoman control to the major trade ports of the eastern Mediterranean. Control of the Mediterranean was essential, if the Ottomans were to implement their economic and political policies in the region.[17] As the Ottoman

---

[14] The earliest reference to *Mahdi* and related terms in the Ottoman Empire is Ahmed Bican Yazıcıoğlu's (d. ca. 1460) cosmographical work *Dürr-i Meknun*. Yazıcıoğlu Ahmed Bican, *Dürr-i Meknun: Saklı İnciler*, trans. Necdet Sakaoğlu (Istanbul: Tarih Vakfı Yurt Yayınları, 1999). For a discussion on the usage of the *Mahdi* in the *Dürr-i Meknun*, see Stéphane Yerasimos, *Légendes d'empire: La fondation de Constantinople et de Sainte-Sophie dans les traditions turques* (Paris: Librairie d'Amérique et d'Orient J. Maisonneuve, 1990); Laban Kaptein, *Eindtijd en Antichrist (Ad-Dağğal) in de Islam: Eschatologie bij Ahmet Bican († ca. 1466)* (Leiden: CNWS, 1997), 275. Celalzade Mustafa, who wrote his *Selimname* during the reign of Süleyman I (r. 1512–20), refers to Selim as *Mahdi* in reference to a poem sent to Istanbul by a poet from Khurasan: "you are a mercy for the reed pen of religion / you are the *Mahdi* of the end of time." Celalzade Mustafa, *Me'asir-i Selim Hani: Selim-name*, ed. Ahmet Uğur and Mustafa Çuhadar (Ankara: Kültür Bakanlığı, 1990), 39–40.

[15] See in particular Selim's contemporaries İshak Üskübi, *Selimname*, Süleymaniye Library, Hekim Ali Paşa 761, 19b; Kemal Paşazade, *Tevarih-i Al-i Osman*, ed. Şefaettin Severcan (Ankara: TTK, 1996), bk. 10, 207 and 209.

[16] Cornell H. Fleischer, "The Lawgiver as Messiah: The Making of the Imperial Image in the Reign of Süleyman," in *Soliman le Magnifique et son temps*, ed. Gilles Veinstein (Paris: La Documentation Française, 1992), 162.

[17] So far, two pioneering historians have analyzed Ottoman political and economic engagement in the sixteenth-century Mediterranean. Andrew Hess, who challenged Braudel's argument about the unity of the Mediterranean, argued that "the separation of the Mediterranean world into different, well-defined cultural spheres [Spanish and Ottoman]" is the main theme of the sixteenth-century Mediterranean and also of the next centuries. Hess emphasized the clash of Islam and Christianity in the early modern Mediterranean and maintained that the chasm grew only wider. Andrew Hess, *The Forgotten Frontier: A History of the Sixteenth-Century Ibero-African Frontier* (Chicago: University of Chicago Press, 1978), 2–3. Recently, Palmira Brummett, who emphasized the idea of common

territories reached the Red Sea, the demographics of the empire changed so that the majority of its population was now Muslim. Prior to these conquests, the ostensibly Muslim Ottoman Empire was comprised mainly of old Byzantine lands and mostly populated with Christian subjects. The changing demographics and geographical positioning of the empire led to the reformulation of its political and economic policies. Provisioning the urban centers and keeping the trade and pilgrimage routes safe became an unequivocal priority for the Ottoman state. This ambitious task required military might in the Mediterranean.

The creation of a stronger Ottoman navy had already started during the reign of Bayezid II (r. 1481–1512).[18] Bayezid increased the size of the fleet and established a close link between the corsairs of the Mediterranean and the imperial fleet.[19] Ottoman sea captains of that era had directly confronted the Knights of Rhodes, the Venetians, and the Spanish Habsburgs in the Mediterranean.[20] After Selim's conquests in the East, the Ottomans controlled Egypt—a major source of grain in the Mediterranean—and the major port cities of Alexandria and Tripoli. The communication lines between Anatolia and Arab lands (especially the holy cities of Mecca and Medina) and the Egyptian markets and ports had to be kept open. As soon as he returned from his eastern campaign, Selim I instructed his grand vizier Piri Mehmed Pasha (d. 1532/33) to enlarge the shipyards in Kasımpaşa

---

space in her analysis, argued that the Ottoman Empire was a formidable power in the early modern Levantine world. Brummett maintained that the Ottoman Empire aimed at control of trade as well as territory in this region. Accordingly, the empire reached these aims through economic and political alliances as well as military and naval force. Palmira Brummett, *Ottoman Seapower and Levantine Diplomacy in the Age of Discovery* (Albany: State University of New York Press, 1994), 8–9.

[18] On Ottoman activities in the Mediterranean as early as the reign of Mehmed II, see Andrew Hess, "The Evolution of the Ottoman Seaborne Empire in the Age of the Oceanic Discoveries 1453–1525," *American Historical Review* 75, no. 7 (1970): 1892–919.

[19] Imber, *The Ottoman Empire*, 40. According to the sixteenth-century chronicler and tutor to Sultan Murad III (r. 1574–95), Hoca Sa'deddin (d. 1599), Bayezid II summoned the corsairs who were active around the coasts and islands of the Aegean to Constantinople and offered them positions within the imperial navy. It was during this period that Kemal Reis (d. 1511) and Burak Reis were appointed as naval captains. Hoca Sa'deddin, *Tacü't-tevarih*, ed. İsmet Parmaksızoğlu (Ankara: Kültür Bakanlığı, 1992), vol. 3, 298.

[20] In the early 1500s, Kemal Reis's fleet consisted of 50 vessels in total. Bayezid had also called in 16 heavy galleys and 40 light galleys anchored in the Black Sea. In 1502, a total of 18 Ottoman ships, which were dispatched from the base in Gallipoli, attacked the island of Leros and several fortresses on Rhodes. The attack did not result in serious damage; however, it confirmed the Ottoman collaboration with the corsairs around the island. Between 1499 and 1503, the Ottomans captured the bays of Lepanto, Navarino, Modon, and Koron along the Peloponnese from the Venetians. Palmira Brummett, "The Ottoman Adversary: Rhodes and Ottoman Naval Power," *Historical Journal* 36, no. 3 (1993): 520–23; Nicholas Vatin, *Rodos* Şövalyeleri ve Osmanlılar: *Doğu Akdeniz'de savaş diplomasi ve korsanlık: 1480–1522*, trans. Tülin Altınova (Istanbul: Türkiye Ekonomik ve Toplumsal Tarih Vakfı, 2004), 255–60.

on the Golden Horn and to construct a new arsenal for the Ottoman navy in Constantinople. Celalzade (d. 1567), a contemporary historian and biographer, describes this event as follows:

> One day they were in Istanbul. There was a council meeting and when the late Pasha came in, they met. And he spread jewels from his words loaded with pearls and declared that there were countries in the climates of the unbelievers in which there were enormous cities, strong and well-protected ramparts and high fortresses, and in the seas there were prosperous islands attracting the heart in endless number. Their rulers were unbelievers. Was it appropriate for unbelievers to sit on the throne and to conquer countries and to rule in this world? Where was the Muslims' zeal? It came to my mind to embark upon preparations with this regard. What is your opinion and suggestion on that, what should be done? The late Pasha was the champion of his time in intelligence and skill, like Bozorgmehr and Asaf in wisdom. He made up his mind and declared that upon the issuance of an imperial rescript, five or six hundred galleys and galliots should be built in the well-protected domains. Upon hearing these words, the sultan praised the Pasha and honored him and said "I hereby order it, start to build them."[21]

As Celalzade tells it, protection of the empire, trade routes, and Islam demanded a strong Ottoman navy. According to both contemporary Ottoman accounts and Venetian reports, Selim I planned to use his new fleet more actively in the Mediterranean. Fittingly, his first naval target was Rhodes, which both his father Bayezid II and grandfather Mehmed II had failed to conquer.[22] In 1519, just as Selim I was turning his attention to the Mediterranean, Hayreddin Barbarossa (d. 1546), a successful corsair in the Mediterranean who had conquered Algiers and other North African sites, entered the Ottoman service. Selim appointed the former corsair governor of Algiers and tried to extend the Ottoman influence in North Africa, especially in Tunis and Algiers.[23] These small but important developments during the reign of Selim contributed to a growing sense in Constantinople that

---

[21] Celalzade Mustafa, *Selim-name*, 45.

[22] Hoca Sa'deddin, *Tacü't-tevarih*, vol. 2, 389; Kenneth M. Setton, *The Papacy and the Levant (1204–1571)* (Philadelphia: American Philosophical Society, 1984), vol. 3, 193.

[23] Hayreddin Barbarossa was engaged in corsair activities around the North African coast in the early sixteenth century with his brother Uruj Reis. In 1510, they made the island Djerba in North Africa their headquarters. By 1516, they had gained control of Algiers. After the death of Uruj Reis in 1518 (and threatened by Spanish advance in the western Mediterranean), Hayreddin Barbarossa joined forces with the Ottoman navy and was appointed governor of Algiers. Seyyid Muradi, *Gazavat-ı Hayreddin Paşa: Barbaros Hayreddin Paşa'nın Hatıraları*, trans. and ed. Ertuğrul Düzdağ (Istanbul: Tercüman Gazetesi, 1980), vol. 1, 210–13. For Barbaros Hayreddin's appointment as the governor of Algiers, see İdris Bostan, "Cezayir-i Bahr-ı Sefid Eyaletinin Kuruluşu (1534)," *İÜEF Tarih Dergisi* 38 (2002–03): 61–77. On the activities of Hayreddin Barbarossa and Uruj Reis on the North African coast, see also İsmail Hakkı Uzunçarşılı, *Osmanlı Tarihi* (Ankara: TTK, 1998), vol. 2, 366–9. On Hayreddin Barbarossa and his influence in North Africa, see Svat Soucek, "The Rise of the Barbarossas in North Africa," *Archivum Ottomanicum* 3 (1972): 228–50.

the navy and its engagements in the Mediterranean were central to the empire's political and economic aspirations. Although Selim's reign ended abruptly in 1520 without a major campaign in the Mediterranean, he left a strong legacy for future Ottoman sultans by strengthening the navy and consolidating control in the eastern Mediterranean.

The Mediterranean, however, was not the only geographic region where the Ottomans had to establish their political existence and influence. After the Hejaz region accepted Selim's suzerainty, the Indian Ocean also entered the empire's political and economic agenda. Historical literature on the Ottomans has long examined Ottoman activities in the Indian Ocean, and the long-term effects of Portuguese trading in the Indies on the Ottoman economy.[24] Recently, Giancarlo Casale argued that Ottoman expansion in the Indian Ocean resembled European expansion patterns during the Age of Exploration, and challenged the Portuguese economic and political network in the region. According to this view, the "Indian Ocean faction," which gained political prestige and prominence in the Ottoman court during the second half of Sultan Süleyman's reign, pursued a conscious agenda to expand Ottoman economic and political power in the Indian Ocean to compete directly with the Portuguese for territory and trade routes.[25]

This book argues that the Indian Ocean faction was not as strong as it has been portrayed and thus the Indian Ocean remained on the "periphery" of the Ottoman imperial project from the first day the Ottomans reached its shores. It is true that contemporary Portuguese sources were steeped in fear of Ottoman expansion in the region and that the battles over trade routes in the Red Sea lent themselves easily to the perception that the Ottomans were fighting to expand as they had done elsewhere. In reality, the Ottomans were very aware of the economic

---

[24] On Ottoman activities in the Indian Ocean, see Andrew Hess, "The Ottoman Conquest of Egypt and the Beginning of the Sixteenth-Century World War," *International Journal of Middle Eastern Studies* 4, no. 1 (1973): 55–76; Yakub Mughul, *Kanuni Devri Osmanlılar'ın Hint Okyanusu Politikası ve Osmanlı-Hint Müslümanları Münasebetleri 1517–1538* (Istanbul: Fetih Yayınevi, 1974); Cengiz Orhonlu, *Osmanlı İmparatorluğu'nun Güney Siyaseti: Habeş Eyaleti* (Istanbul: TTK, 1974). A pioneer in this field is Salih Özbaran, who used both Portuguese and Ottoman archival documents: Salih Özbaran, "Osmanlı İmparatorluğu ve Hindistan Yolu: Onaltıncı Yüzyılda Ticaret Yolları Üzerinde Türk-Portekiz Rekabet ve İlişkileri," *Tarih Dergisi* 31 (1977): 66–146. For a collection of Özbaran's articles, see Özbaran, *The Ottoman Response to European Expansion: Studies on Ottoman-Portuguese Relations in the Indian Ocean and Ottoman Administration in the Arab Lands during the Sixteenth Century* (Istanbul: Isis, 1994) and *Ottoman Expansion towards the Indian Ocean in the 16th Century* (Istanbul: Bilgi University, 2009). On the Ottoman Empire as an active participant in the contest for commercial and political hegemony in the Indian Ocean, see Halil İnalcık, *An Economic and Social History of the Ottoman Empire* (Cambridge: Cambridge University Press, 1994), vol. 1, 315–64; Palmira Brummett, *Ottoman Seapower and Levantine Diplomacy in the Age of Discovery* (Albany: State University of New York Press, 1994); Giancarlo Casale, *The Ottoman Age of Exploration* (Oxford: Oxford University Press, 2010).

[25] Casale, *The Ottoman Age of Exploration*, 84–116.

importance of the Indian Ocean and its significant role in the spice trade.[26] That said, contemporary Ottoman accounts of events in the Indian Ocean exhibit none of the urgency or fear that Ottoman sources use to depict conflicts and campaigns in the Mediterranean.

Selim's eastern campaigns and later the control of Egypt and the Hejaz region gave the Ottomans easy access to the Red Sea. Already during the Mamluk rule in Egypt, Selman Reis, a former Mediterranean corsair and an Ottoman sea captain, had already overseen the construction of a fleet in the Red Sea in order to stop Portuguese navigation in the area.[27] After the Portuguese viceroy Alfonso de Albuquerque took control of Kamaran Island in 1515, the Portuguese navy became an open threat to Muslim ships in the Red Sea.[28] Complaints about the Portuguese reached Topkapı Palace during the reign of Selim I, and the sultan's advisers urged him to take action.[29] By the time the Ottomans conquered Egypt in 1517, the Portuguese navy had already reached the west coast of India and threatened merchant ships sailing between the Red Sea and Malabar in western India.[30] Selim, who acknowledged the problems posed by the Portuguese presence in the Indian Ocean and their impact on Muslim spice traders, endeavored to stop Portuguese expansion in the area.[31] This, however, did not imply an intention to expand the Ottomans' territory in the region, but rather reflected a defensive posture—to keep the empire free from threats. For the Ottoman court under Selim I, the Indian Ocean was simply not as important as the other parts of the Ottoman realm. It never served as an arena for the great battles between the sultan and Christian or Shi'a rulers. For Selim and his ruling elites, the Red Sea and the Persian Gulf were important hinterlands of the Ottoman domains, but were not objects of conquest. Instead, they were secured with an eye to defense. The Red Sea was vital for the protection of the Hejaz region and the Holy Cities, while the Persian Gulf could threaten the Safavids with a second front.

A Renaissance of Ottoman Geographical Consciousness

This book argues that as Selim's military campaigns and conquests laid the foundations of a nascent empire, the Ottoman state translated the empire's changing cultural, geographic, and demographic structure into a new imperial outlook.

26 Ibid., 13–33.

27 İdris Bostan, *Osmanlı Bahriye Teşkilatı: XVII. Yüzyılda Tersane-i Amire* (Ankara: TTK, 1992), 20. According to İdris Bostan, the Ottomans also started building tall-sided ships similar to those used by the Portuguese in the Indian Ocean to support the navy: Bostan, *Osmanlılar ve Deniz: Deniz Politikaları, Teşkilat, Gemiler* (Istanbul: Küre, 2007).

28 For Portuguese activities in the Red Sea, see especially Salih Özbaran, "The Portuguese in the Red Sea," in *The Ottoman Response to European Expansion*, 89–97; Casale, *The Ottoman Age of Exploration*, 13–33.

29 See especially TKS. E. 8009; E. 6472.

30 Özbaran, "Portuguese in the Red Sea," 93.

31 Casale, *The Ottoman Age of Exploration*, 28–9.

In this new vision, the Ottoman sultan now embodied the role of world conqueror and universal sovereign. In the leading Ottoman political ideology, universal sovereignty was intimately related with the control of the Mediterranean and its trade and travel routes. In order to substantiate these claims, the Ottoman state, during the reign of Selim, took control of the major ports in the eastern Mediterranean and strengthened its military power to challenge the Venetians and the Habsburgs in the central and western Mediterranean. At the same time, Selim and his ruling elites also welcomed and encouraged intellectuals to articulate these claims in their works. Among these intellectuals were a group of geographers who provided both symbolic and practical means to validate this new imperial vision.[32]

As the Ottoman court took advantage of geographical accounts, Ottoman geographers created a unique Ottoman tradition by combining pre-Ottoman Muslim and European conventions and practices. The earliest Ottoman geographical accounts were in fact fourteenth- and fifteenth-century translations of Arabic cosmographical works.[33] Translations of Zakariya al-Qazwini's (d. 1283) *Aja'ib al-Mahluqat* (*Wonders of Creation*) and Ibn al-Wardi's (d. 1349) *Kharidat al-'Aja'ib wa Faridat al-Ghara'ib* (*Pearls of Marvels and Unique Curiosities*) were among the first works on geography in the Ottoman language.[34] Most well known was Yazıcıoğlu Ahmet Bican's (d. 1456) abridged translation of al-Qazwini's *Aja'ib al-Mahluqat*. Yazıcıoğlu also wrote the *Dürr-i Meknun* (*Well-preserved Pearl*), which was an example of a small cosmography and geography in the style of *Pearls of Marvels*—a treatment of the wonders of Creation.[35]

---

[32] On the relationship between state policies and developments in the field of geography in early modern European empires, see David Livingstone, *The Geographical Tradition: Episodes in the History of a Contested Enterprise* (Oxford: Blackwell Publishing, 1992), 32–102; Frank Lestringant, *Mapping the Renaissance World: The Geographical Imagination in the Age of Discovery*, trans. David Fausett (Cambridge: Polity Press, 1994); David Woodward, "Cartography and the Renaissance: Continuity and Change," in *The History of Cartography, Vol. 3, Bk. 1, Cartography in the European Renaissance*, ed. J. B. Harley and David Woodward (Chicago: University of Chicago Press, 2007), 3–24. For the Spanish case in particular, see Ricardo Padrón, *The Spacious Word: Cartography, Literature, and Empire in Early Modern Spain* (Chicago: University of Chicago Press, 2004); Maria M. Portuondo, *Secret Science: Spanish Cosmography and the New World* (Chicago: University of Chicago Press, 2009).

[33] Ekmeleddin İhsanoğlu, *Osmanlı Coğrafya Literatürü Tarihi* (Istanbul: IRCICA, 2000), vol. 1, xxxvi; Franz Taeschner, "Osmanlilarda Coğrafya," *Türkiyat Mecmuasi* 2 (1926): 271–314; Günay Kut, *Acaibü'l-Mahlukat: Eski Türk Edebiyatı Araştırmaları II*, ed. Fatma Büyükkarcı Yılmaz (Istanbul: Simurg Kitapçılık, 2010), 1–11.

[34] 'Ali b. Abdurrahman's *'Aca'ib al-Mahlukat* is considered the oldest Ottoman work on cosmography, even though the earliest and incomplete manuscript is from the seventeenth century. Gottfried Hagen, *Ein osmanischer Geographer bei der Arbeit Entstehung und Gedankenwelt von Katib Celebis Cihannüma* (Berlin: Klaus Schwarz Verlag, 2003), 85.

[35] Franz Taeschner, "Djughrafiya: The Ottoman Geographers," *EI²*, last accessed via Brill Online. University of Pittsburgh. 7 November 2011; Cevdet Türkay, *Osmanlı Türklerinde Coğrafya* (Istanbul: Maarif Basımevi, 1959), 14.

The influence of Islamic traditions was not, however, limited to these translations. Compilations integrating various Islamic geographical works, such as Sipahizade's (d. 1589) historical geography in Arabic, *Awdah al-Masalik fi Ma'rifat al-Buldan wa'l-Mamalik* (*Explanation of the Routes about the Knowledge of Cities and Countries*) and Aşık Mehmed b. Ömer's (d. after 1605) cosmography *Menazirü'l Avalim* (*Views of the World*), were common endeavors among Ottoman intellectuals in the early modern period.[36] Further, inspired by *ziyarat* literature, which consisted of works describing towns and places of religious significance or places of pilgrimage, Ottoman intellectuals prepared similar accounts. In these works, one finds descriptions of the length of the road, either in time or in mileage, from Istanbul, Cairo, or Damascus to Mecca, together with descriptions of important places along the way. Shams al-Din Muhammad's sixteenth-century *Manazil al-Haramayn* (*Stations of the Two Holy Places*) in Arabic is the oldest example of this genre within the Ottoman lands.[37] Ottomans likewise produced their own examples of travel literature, such as campaign diaries with descriptions of important cities and towns along campaign routes. Matrakçı Nasuh's *Mecmu'-i Menazil: Beyan-ı Sefer-i 'Iraqeyn-i Sultan Süleyman Han* is the most studied and appreciated work of this genre.[38]

It was during the reign of Mehmed II (r. 1444–46, 1451–81), who himself collected and drew maps, that geographical accounts increasingly circulated in Constantinople. Mehmed was also known to have commissioned Byzantine and European geographers to translate the works of Ptolemy and to prepare atlases for him.[39] It was during the reign of Selim I, however, that Ottoman geographers endeavored to consolidate Ottoman involvement in geography and geographical understanding. Their aim was to turn the Ottoman Empire into an active player in the transmission of geographical knowledge across the Mediterranean. Simultaneously, they strove to earn the favor of the Ottoman court by displaying their command of latest knowledge. In doing so, they introduced new geographic knowledge in a way that was easily read and visualized and in doing so created a more accessible body of knowledge in Ottoman Turkish. In their works, they combined knowledge garnered from travel accounts and maps circulating in the Mediterranean with their own experiences. This does not mean that they ignored information readily

---

36 İhsanoğlu, *Osmanlı Coğrafya Literatürü Tarihi*, vol. 1, xxxvii. Both these works are valuable contributions to the field of geography in the sixteenth-century Ottoman Empire. As opposed to the works that are the main focus of this book, however, they hardly present new knowledge attained through personal experience or from European accounts. For instance, Aşık Mehmed b. Ömer's encyclopedic work on universal cosmography, *Menazirü'l Avalim*, is a compilation of traditional Muslim accounts. On Aşık Mehmed b. Ömer and his universal cosmography, see Aşık Mehmed, *Menazirü'l-Avalim*, ed. Mahmut Ak (Ankara: TTK, 2007), vol. 1, xxxvii–cci.

37 Ibid., xxxviii.

38 For the historical literature on Matrakçı Nasuh and his works, see footnote 112.

39 Brotton, *Trading Territories: Mapping the Early Modern World* (Ithaca: Cornell University Press, 1997), 87–119.

available in Muslim accounts of cosmography and universal geography. However, their emphasis was on new knowledge articulated in both old and new styles.[40]

In 1517, while Selim I was in Cairo, Ottoman sea captain and geographer Piri Reis (d. 1554) presented him with a world map—a map that he had finished in 1513.[41] Piri Reis was an ambitious political figure who sought the patronage of Selim I. By presenting his world map to Selim, he had one major aim: to promote himself before the sultan by displaying his geographical knowledge and thereby gaining Selim's support. Unfortunately, the eastern two-thirds of the map have been lost. The extant section depicts Africa, the western coast of Europe, the Atlantic Ocean, and the eastern coast of the New World. On the map, there is a long legend in which Piri Reis lists his sources: 20 *portolan*s (sea charts), a world map, eight Arab and four Portuguese sailing charts, and a map created by Columbus.[42] In his own words, Piri Reis "took into consideration the maps that have recently appeared of the seas of India and China that up to now nobody knew in the lands of Rum."[43] Since the sections of the map depicting the Mediterranean, China, and the Indian Ocean are missing, we cannot easily validate Piri Reis's claims. Today, the manuscript libraries in Istanbul hold copies of the *portolan* charts by al-Katibi from Tunis (1413–14), the Majorcan cartographer Johannes de Villadestes (1428), and İbrahim al-Mursi (1461), and a world map made by Pedro Reinel (ca. 1519).[44] It is possible that Piri Reis drew on these and other maps. As a sea captain, he had easy access to other sea charts circulating in the Mediterranean. Piri Reis entered the Ottoman service as a sea captain at the end of the fifteenth century. His stint as a corsair and later as a captain in the Ottoman navy gave him the opportunity to navigate the Mediterranean and to learn about the problems this body of water could present. As a corsair, he must have familiarized himself with the languages used in the Mediterranean—mainly Italian, the *lingua franca* of the sixteenth century, Greek, and perhaps Spanish. This might have enabled him to decipher and understand the maps he came upon while sailing around the Mediterranean.

---

[40] In this period, there is clearly a tension between the old knowledge represented in more traditionally oriented geographical works and the new knowledge articulated in new forms and styles. However, this tension is beyond the scope of this work. Here, the focus is on the geographers who chose to present the new knowledge that they acquired by their own experiences in new and old ways.

[41] On Piri Reis's world map and his life, see A. Afetinan, *Life and Works of Piri Reis* (Ankara: TTK, 1987); A. Adnan Adıvar, *Osmanlı Türklerinde İlim*, 5th ed. (Istanbul: Remzi Kitabevi, 1991), 74–6; Svat Soucek, "Piri Reis," *EI*[2], last accessed via Brill Online. University of Pittsburgh. 7 November 2011.

[42] Piri Reis, *Mappamundi*, TKS. Revan Köşkü 1633 mük.

[43] Piri Reis, *Kitab-ı Bahriye*, Süleymaniye Library, Ayasofya 2612, vol. 1, 3a.

[44] Svat Soucek, "Islamic Charting in the Mediterranean," in *The History of Cartography, Vol. 2, Bk. 1, Cartography in the Traditional Islamic and South Asian Societies*, ed. J. B. Harley and David Woodward (Chicago: University of Chicago Press, 1992), 264–6; İhsanoğlu, *Osmanlı Coğrafya Literatürü Tarihi*, vol. 1, 3–4.

Piri Reis was a pioneer of geographical literature in the Ottoman Empire. In this world map, he advanced his own vision of what a good geographical work should encompass: the latest knowledge. The idea of integrating the latest knowledge into one's work was in fact new both for Ottoman and European geographers. Instead of relying entirely on the knowledge of the ancients, sixteenth-century geographers both in Europe and the Ottoman Empire frequently started to indicate the necessity of integrating the latest observations and discoveries into their works. In European history, these discussions are usually treated as signs of the European Renaissance.[45] Traditional historiography on the Renaissance usually omits the Ottomans from such discussions and portrays them as completely ignorant of these developments and as passive viewers of Renaissance ideals.[46] In the past decades, historians have attempted to broaden the definition of "Renaissance" to include non-European parts of the world in their work.[47] Even so, the Ottomans are usually either completely omitted or regarded as influencing the "European Renaissance."[48] This book maintains that the early sixteenth century was the period of the Ottoman renaissance in geographical knowledge. The Ottoman geographers were aware of the latest knowledge circulating in the Mediterranean. In fact, Piri Reis and his contemporary Ottoman geographers espoused renaissance ideals and were active participants in discussions of "recent knowledge." Piri Reis was the first Ottoman geographer who openly advocated integrating the latest knowledge in geographical accounts rather than depending on traditional Muslim accounts or Ptolemy's *Geographia*.

The early sixteenth century was the time when the Ottomans embarked upon territorial expansion in the Mediterranean and Southeastern Europe. As the geographical and demographical structure of the empire changed, the Ottoman

---

[45] See Anthony Grafton, *New Worlds, Ancient Texts: The Power of Tradition and the Shock of Discovery* (Cambridge, MA: Belknap Press, 1992); John Jeffries Martin, "The Renaissance: A World in Motion," in *The Renaissance World*, ed. John Jeffries Martin (New York and London: Routledge, 2007), 3–31; Francisco Bethencourt, "European Expansion and the New Order of Knowledge," in *The Renaissance World*, ed. John Jeffries Martin, 118–40.

[46] Jacob Burckhardt, *The Civilization of the Renaissance in Italy* (1860; New York: Modern Library, 2002).

[47] See especially Jerry Brotton, *The Renaissance Bazaar: From the Silk Road to Michelangelo* (Oxford: Oxford University Press, 2002); Linda Darling, "The Renaissance and the Middle East," in *A Companion to the Worlds of the Renaissance*, ed. Guido Ruggiero (London: Blackwell Publishers, 2002), 55–87; Nancy Bisaha, *Creating East and West: Renaissance Humanism and the Ottoman Turks* (Pittsburgh: University of Pennsylvania Press, 2004); Daniel Goffman, "The Ottoman Empire," in *The Renaissance World*, ed. John Jeffries Martin, 347–64; Jack Goody, *Renaissances: The One or the Many?* (Cambridge: Cambridge University Press, 2010).

[48] Jack Goody, who argues for a multiplicity of cultural rebirths in different times and locations throughout world history, almost completely omits the Ottomans from his analysis. Goody, *Renaissances*, 94–144. Jerry Brotton regards the Ottoman sultans, especially Mehmed II, as active members of the "European Renaissance" as patrons of the arts and geographical works. Jerry Brotton, *The Renaissance Bazaar*, 48–54.

sultans welcomed artists and intellectuals to articulate the new outlook of the empire. There was no school of navigation in Constantinople similar to the *Casa de la Contratación* (House of Trade) in Spain. The *Casa*, established in Seville in 1503, was charged with regulating all Spanish commerce and navigation to the New World and to the Indies. It recruited skilled cartographers and sea pilots to train navigators and to draw maps of the Habsburg possessions in the New World. Under Charles V and his son Philip II, the institution became an important center for organizing and controlling the production and distribution of cartographical works. The *Casa* did not promote geographical knowledge for its own sake, however. In fact, one of the most important missions of the institution was to censor cartographic knowledge.[49] In the Ottoman Empire, the task of organizing geographic knowledge and enhancing the Ottoman geographic understanding fell to individual cartographers and geographers.

Who were these geographers then? Who were their patrons? How did they earn their livelihood? This book maintains that these geographers were Ottoman intellectuals who were either already part of the Ottoman administration such as Matrakçı Nasuh or aspired to be part of the system like Mustafa b. Ali by earning the Ottoman court's patronage. Among them were seamen who tried to earn promotion in the Ottoman navy by articulating the claims of one or the other faction in Constantinople. For instance, both Piri Reis and Seydi Ali Reis came from the ranks of seafarers. Therefore it is safe to argue that geography was not the primary occupation for these officers and intellectuals. However, they employed it to earn the patronage of the leading political figures of the Ottoman court. In doing so, they articulated the political agendas of their patrons and patrons-to-be by representing new knowledge through old and novel means. Furthermore, they pioneered the introduction of geography as a distinct field, divorced from history, into the intellectual arena.

Piri Reis was as close to an official cartographer as one could get in the Ottoman Empire.[50] He and his works epitomized a new era of geographical understanding in Ottoman history where geographical knowledge and imperial culture were closely woven together. In the early sixteenth century, cartographers and mapmakers who drew and commissioned maps for important political figures in Europe tried to incorporate the latest geographical discoveries into their works.[51] Henricus Martellus Germanus's world map (1488) and the Nuremberg merchant Martin Behaim's terrestrial globe (1492) are the earliest examples of cartographical works of this sort. These cartographers endeavored to include the latest findings in the known forms of Ptolemaic maps and *portolan* charts. Behaim and Martellus

---

49 Padrón, *The Spacious Word*, 8; Portuondo, *Secret Science*, 95–102.

50 Svat Soucek, "Piri Reis and Ottoman Discovery of the Great Discoveries," *Studia Islamica* 79 (1994): 134–5.

51 Felipe Fernandez-Armesto, "Maps and Exploration in the Sixteenth and Early Seventeenth Centuries," in *The History of Cartography, Vol. 3, Bk. 1, Cartography in the European Renaissance*, ed. Harley and Woodward, 754–8.

Germanus both derived information on India, the Indian Ocean, and Madagascar from Marco Polo's accounts.[52] These maps and many others, prepared in the same period in different parts of Europe, were drawn up with two aims in mind: to please the cartographers' patrons, and to relate the most recent information on the geography of the world.

Piri Reis's aims and ambitions were not different from other contemporary cartographers'. He presented his work to the Ottoman sultan, whom he perceived as a prestigious patron. In doing so, he not only enhanced the Ottoman sultan and his court's vision of the world but also buttressed its imperial claims by providing practical knowledge that could assist Ottoman armies during imperial battles. As two-thirds of the map is missing, we do not know how Piri Reis projected the Indian Ocean. The geographer later composed another valuable geographical work, a navigational guide for Selim's son, Süleyman I (r. 1520–66). In this work, he focused exclusively on the Mediterranean. This first *isolario* in Ottoman Turkish—together with his world map—further contributed to the development of geography as a separate genre in the Ottoman Empire.

Süleyman the Magnificent and the Ottoman "Grand Project"

> He [Süleyman] is the hero of his age and became second to Sam, a third Persian chief like Rustam-e Esfandiyar, in the arena of bravery, making the rulers of four corners of the inhabited quarters of the world suffer under his tormenting hand, and by his justice, equity and munificence becoming an example of the eight heavens of glory in the six universes of the seven climes of the world.[53]

When Selim I died suddenly of natural causes in 1520, the only candidate for the Ottoman throne was his son Süleyman. Süleyman I peacefully inherited an empire that encompassed lands extending from the Arabian Peninsula in the east to the Balkans in the west to Egypt in the south. He assumed a host of imperial titles, including that of the "sultan of seven climes," a direct and geographical reference to the universal sovereignty of Ottoman sultans. At the onset of Süleyman's reign, the fragile young sultan had two important tasks. First, because he was still surrounded by the aides and advisers of his deceased father, he had to prove himself as a capable ruler. Süleyman's peaceful ascension to the throne was actually a marked disadvantage. Ottoman tradition pitted all eligible sons

---

52 David Woodward, "Medieval Mappaemundi," in the *History of Cartography, Vol. 1, Cartography in Prehistoric, Ancient, and Medieval Europe and the Mediterranean*, ed. J. B. Harley and David Woodward (Chicago: University of Chicago Press, 1987), 316. Piri Reis's world map is drawn in the tradition of *portolan* charts and is contemporaneous with the Cantino map (1502), the Pesaro map (ca. 1505–08), and Vesconte Maggiolo's map (1516).

53 Süleyman's chancellor Ramazanzade's description of his patron (ca. 1553–54). Ramazanzade Nişancı Mehmed Pasha, *Tarih-i Nişancı Mehmed Paşa* (Istanbul, 1290/1873), 197.

against one another, each with equal claim to the throne. This tradition finds its roots in Central Asia where properties were split evenly among sons. In Europe, by contrast, properties were traditionally bequeathed to the eldest son. Because the empire was a prize that could not be divided, previous Ottoman heirs had engaged in a very visible struggle for the throne by demonstrating their worth in public as soldiers and rulers.[54] Süleyman's unchallenged succession, therefore, did not impress observers.

In addition to proving himself to aides and advisers, Süleyman was also confronted with the task of consolidating the imperial claims to universal sovereignty that had evolved so rapidly during the reign of his ambitious father. Selim's repute as a "ruthless" conqueror circulated widely through European courts.[55] When he died, Europe breathed a collective sigh of relief. The young Süleyman, by comparison, was viewed as a "peaceful lamb," giving many Europeans hope that Selim's death marked an end to conquest and universal sovereignty.[56]

Around the time that Süleyman ascended to the Ottoman throne, Europe witnessed the rise of Charles V, a charismatic and bold leader who strove to satisfy the Spanish and Austrian/Germanic aspirations of creating a universal monarchy. His advisers touted him as the "Last World Emperor," capable of joining together the territories of the Roman Empire and the New World. This image spread widely across the German-speaking parts of his domain. He was portrayed as the emperor who would restore the universal claims of the Holy Roman Empire.[57] In 1519, Charles added Austria to his domains, which already included Spain, Naples, the Netherlands, and parts of the New World. The same year, he was elected Holy Roman Emperor. Süleyman and Charles, each ambitious and intelligent, soon became famous rivals. Their militaries fought in Hungary and the Mediterranean, while their palaces saw advisers and intellectuals articulating competing imperial discourses.[58]

---

54 Leslie Peirce, *The Imperial Harem: Women and Sovereignty in the Ottoman Empire* (New York: Oxford University Press, 1993), 25.

55 Ebru Turan, "The Sultan's Favorite: Ibrahim Pasha and the Making of the Ottoman Universal Sovereignty in the Reign of Sultan Süleyman (1516–1526)" (PhD diss., University of Chicago, 2007), 61–3.

56 Ibid., 64.

57 Frances A. Yates, *Astrea: The Imperial Theme in the Sixteenth Century* (London: Ark, 1985), 20–21; Peter Burke, "Presenting and Re-presenting Charles V," in *Charles V, 1500–1558, and His Time*, ed. Hugo Soly (Antwerp: Mercatorfonds, 1999), 411–33.

58 For Charles's and Süleyman's reciprocal claims for universality, see Gülru Necipoğlu, "Süleyman the Magnificent and the Representation of Power in the Context of Ottoman-Hapsburg-Papal Rivalry," *Art Bulletin* 71, no. 3 (1989): 401–27; Robert Finlay, "Prophecy and Politics in Istanbul: Charles V, Sultan Süleyman, and the Habsburg Embassy of 1533–1534," *Journal of Early Modern History* 2, no. 1 (1998): 1–31; Ágoston, "Information, Ideology, and Limits of Imperial Policy," 96–100.

Süleyman's need for an auspicious start to his reign took him first to Belgrade.[59] He seized the city in 1521 and then conquered Rhodes in 1522. His illustrious great-grandfather, Mehmed II, the conqueror of Constantinople, had failed in both places. Writing around 1540, Bostan Çelebi, a chronicler during Süleyman's reign, describes the importance of the conquest of Rhodes for his patron and the Muslim world as follows:

> The enforcing power and obvious might of His Majesty the ruler over the world, may God exalt his victories, had reached such a degree that the surface of the lands of Islam and the whole of mankind was purified tip to toe from the filth and attacks of the enemies of religion and state so that among the rulers of the inhabited quarters of the world no individual would breathe any hint of opposition or intend on confrontation, and the great cities were surrendered and conquered. But the unbelievers of the island of Rhodes in the Mediterranean, who trusted in the strength of their fortifications and relied on their ultimate inaccessibility, indulged in piracy all over the sea, crossed the routes of the Muslims, took their ships and, from across the sea, spread disorder on the coasts of the Islamic lands as well, spilling blood and robbing property. How many strong rulers from among the Islamic sultans had equipped one army after another without winning a victory that would permit conquest, so that from the origins of our religion to these times nobody was capable of conquering and pacifying Rhodes, which remained an untouched virgin.[60]

Süleyman's early conquests in Belgrade and Rhodes brought him new lands and greater prestige. They also provided legitimization in the eyes of palace dignitaries, most of whom had served under Selim I.[61]

After the conquests of Belgrade and Rhodes, Süleyman I became an open threat to the Holy Roman Emperors. In this period, both Charles V and Ferdinand I prophesied the destruction of the Ottoman Empire and the reconquest of Constantinople.[62] Especially after Selim I's conquests between 1514 and 1517,

---

59 The new sultan started his reign by triumphing where his glorious ancestors had failed. Belgrade had been an important target for Ottoman rulers since the reign of Mehmed II. It was a strategic location to control Hungary and thus the whole Balkans. In 1456, Mehmed II attempted, but failed, to conquer the city. For the details of Mehmed II's preparations for the siege, see Franz Babinger, *Mehmed the Conqueror and His Time*, trans. Ralph Manheim, ed. William Hickman (Princeton: Princeton University Press, 1978), 138–50. This failure added ideological importance to the conquest of Belgrade. Since Süleyman ascended the throne without any rivalry, Belgrade was the first occasion for the young sultan to prove himself. Ramazanzade, *Tarih-i Nişancı Mehmed Paşa*, 216–18; Bostan Çelebi, *Gazavat-ı Sultan Süleyman*, Süleymaniye Library, Ayasofya 3117, 19b–39a.

60 Bostan Çelebi, *Gazavat-ı Sultan Süleyman*, 38b–39a.

61 Turan, "The Sultan's Favorite," 72–99.

62 Kenneth M. Setton, *Western Hostility to Islam and Prophecies of Turkish Doom* (Philadelphia: American Philosophical Society, 1992), 16–17. For further examples of the Turkish image in the eyes of contemporary Europeans, see John S. Geary, "Arredondo's *Castillo inexpugnable de la fee*: Anti-Islamic Propaganda in the Age of Charles V," in *Medieval Christian Perceptions of Islam*, ed. John Victor Tolan (New York: Garland, 1996), 291–311.

European powers identified Islam with the Ottoman Empire, and plans for organizing a crusade against the Ottomans were made in the palaces of the emperors and the pope. The death of Selim I in 1520, however, relieved Europe. Thus, Europe focused on the conflict between François I of France and Charles V. In this period, Lutheranism in Europe replaced the Ottoman menace.[63] After his coronation as Holy Roman Emperor in 1519, Charles V adopted the role of crusader and savior of Christianity. The prophecies written after his coronation "foretold his destruction of the Ottoman Empire as well as his conquest of most of Europe."[64] Similar prophecies also depicted Charles V as champion of the Church against the Lutherans.[65] However, after Süleyman I's successes in Belgrade and Rhodes, Charles V's and the Pope's attention was again directed toward the Ottomans.[66] In this rivalry, the alliances of Süleyman I with François I of France, his warmer attitude toward the Lutherans, as well as the trade connections with Venice, articulated the Muslim Ottoman stance against the Habsburg Empire.[67]

After the conquest of Belgrade and Rhodes, Süleyman also felt emboldened enough to begin making important changes to the palace staff. He replaced his father's grand vizier with his "favorite," İbrahim Ağa, who was then the head of the Privy Chamber. While İbrahim Ağa, a personal servant of Süleyman, had no experience in government, his outsider status proved to be a blessing for Süleyman. İbrahim Pasha, as he came to be known, was able to provide unbiased advice upon which the sultan could rely.[68] Until his tragic execution in 1536, İbrahim Pasha was a friend and true confidant for Süleyman; he directed the sultan's imperial policies and political strategies. Süleyman granted İbrahim Pasha unprecedented power, almost equal to his own. At the height of his powers, İbrahim Pasha became a direct extension of the sultan's will and personality.[69]

---

63 Kenneth M. Setton, "Pope Leo X and the Turkish Perils," in *Europe and the Levant in the Middle Ages and the Renaissance* (London: Variorum Reprints, 1974), 419–20.

64 Setton, *Western Hostility to Islam*, 24.

65 Ibid.

66 Indeed, Luther also preached for a united front against the Ottoman Empire. According to Luther, uniting this front was not incumbent upon the Pope, but on the European states. K. M. Setton, "Lutheranism and the Turkish Peril," in *Europe and the Levant in the Middle Ages and the Renaissance*, 141–8.

67 Cemal Kafadar, "The Ottomans and Europe," in *Structures and Assertions*, vol. 1 of *Handbook of European History 1400–1600: Late Middle Ages, Renaissance and Reformation*, ed. Thomas A. Brady, Jr., Heiko A. Oberman, and James D. Tracy (Leiden: E. J. Brill, 1994), 609.

68 Turan, "The Sultan's Favorite," 104.

69 Following the military failures in Vienna in 1529 and 1532 and the successful campaign against the Safavids from 1533 to 1536, Süleyman sacrificed his "favorite" grand vizier and had him executed. According to the historical literature, among the reasons for this execution were İbrahim Pasha's ambitions, which exceeded the space granted to him by the Sultan, and Süleyman's effort to restore the boundaries of his power. Peirce, *The Imperial Harem*, 72–5.

İbrahim Pasha and the Consolidation of the Imperial Enterprise

İbrahim Pasha eventually assumed near-complete stewardship of day-to-day operations of the empire, while the office of the sultan witnessed a change in protocol. İbrahim Pasha altered the ceremonial rules and regulations of the palace so that the sultan no longer had routine contact with the public.[70] In the past, Ottoman sultans occasionally had informal contact with their subjects during festivals and public ceremonies, while his advisers and viziers could engage the sultan in debate and conversation. With İbrahim's change of rules, this direct link between the sultan and the public was severed.

Under İbrahim Pasha's leadership, Ottoman military campaigns and political discourse sent one clear message: Sultan Süleyman was challenging the authority of both the pope and the Holy Roman Emperor. The sultan was the new world conqueror, the expected Last Emperor, a universal ruler who would unite all mankind under one leader and one religion before the Day of Judgment.[71] While Charles V and his advisers had made similar claims, the Ottoman ruling elites had one big advantage: Süleyman assumed that he was the rightful universal ruler, the successor of his favorite historical hero, Alexander the Great.[72] In accordance with the apocalyptic expectations of his age, the tenth century of the Muslim era, Süleyman I also proclaimed that he would disseminate justice around the world as the messianic ruler.[73]

Although these claims bear similarities to those of Selim, Süleyman's advisers had an easier time of portraying the Ottoman sultan as the world conqueror and defender of Islam. Especially during the first half of Süleyman's reign, most Ottoman military campaigns challenged the empire's Christian enemies, particularly in Charles's domains in Europe. During this period, the Ottoman-Habsburg rivalry focused specifically on control of Hungary.[74] The Ottomans

---

[70] Gülru Necipoğlu, *Architecture, Ceremonial, and Power: The Topkapı Palace in the Fifteenth and Sixteenth Centuries* (New York: Architectural History Foundation, 1991), 22–5.

[71] For discussions on global strategic policies and universal claims in the early modern world, see Geoffrey Parker, *The Grand Strategy of Philip II* (New Haven: Yale University Press, 1998); John P. LeDonne, *The Grand Strategy of the Russian Empire, 1650–1831* (Oxford: Oxford University Press, 2004). For the Ottoman case, see especially Fleischer, "The Lawgiver as Messiah"; Ágoston, "Information, Ideology, and Limits of Imperial Power"; and Necipoğlu, "Süleyman the Magnificent and the Representation of Power."

[72] Finlay, "Prophecy and Politics in Istanbul," 22; Ágoston, "Information, Ideology, and Limits of Imperial Power," 99.

[73] Fleischer, "The Lawgiver as Messiah," 159–65.

[74] On the Ottoman campaign in Hungary and Ottoman policies in Hungary, see Pál Fodor, "Ottoman Policy towards Hungary," *Acta Orientalia Academiae Scientiarum Hungaricae* 45 (1991): 271–345; Tayyib Gökbilgin, "Kanunî Sultan Süleyman'ın Macaristan ve Avrupa Siyasetinin Sebep ve Âmilleri, Geçirdiği Safhalar," in *Kanuni Armağanı* (Ankara: TTK, 1970), 12–16. Among the reasons for the Hungarian campaign

entered Budin in 1526 and claimed Hungary as an Ottoman vassal state. As Charles V's forces sacked Rome in 1527, Süleyman I was planning his second venture to Hungary. The sultan's second campaign to Hungary in 1529 targeted Vienna, the royal capital of the Habsburgs. Charles V, who at the time was busy pacifying the Italian Peninsula, had left the defense of the city to his brother Ferdinand.[75] Süleyman's long siege ended in failure in 1529, but his armies entered Hungary again in 1532.

İbrahim Pasha, mastermind of the failed campaigns in Hungary, came under criticism after a treaty in 1533 left western and northern Hungary in Habsburg control, prompting him to orchestrate a new campaign to restore faith and confidence among onlookers. The grand vizier carefully scripted Süleyman's triumphal entries into Niš and Belgrade, spectacles designed for Western observers and European diplomats based in these cities. During the massive processions, Süleyman carried a tall helmet made of gems. It also featured four superimposed crowns and was topped with a tiara. Commissioned by the grand vizier İbrahim Pasha, the priceless helmet was modeled after the crown that Charles V wore during his coronation celebration in Bologna in 1530, and also the papal tiara.[76] Ottoman sultans had not worn or carried crowns before (and did not wear them after) the march into Southeastern Europe. The symbolic meaning of the object was clear. Messianic ambitions aside, the campaigns in Hungary had been expensive and exhausting disappointments, failing to result in territorial and economic gains. Through pomp and pageantry, the Ottomans were attempting to reassert their superiority.

With calm restored in Southern Europe, the Ottoman-Habsburg rivalry entered a new phase in a new venue—the Mediterranean. Two naval captains tried to claim the sea for their respective imperial patrons: Andrea Doria, the Genoese admiral of a Habsburg armada, and Hayreddin Barbarossa, the admiral of the Ottoman fleet. The two often challenged each other, especially in the Aegean Sea and North Africa, where the Ottomans and Habsburgs vied for control of the strategically important city of Tunis. In 1533, after the Ottoman-Habsburg conflict for control of Hungary ended in a stalemate, Andrea Doria, the Genoese admiral of a Habsburg armada, challenged the Ottoman fleet near the Peloponnese and seized the ports of Naupactus and Coron. In response, Sultan Süleyman appointed Hayreddin

---

was to help François I, who had been imprisoned by Charles V. See Kemal Paşazade, *Tevarih-i Al-i Osman*, bk. 10, 23a–23b; Matrakçı Nasuh, *Süleymanname*, TKS. Revan Köşkü 1286, 99a–99b. Here, Matrakçı Nasuh replicates Kemal Paşazade's account word for word. In fact, Matrakçı copies the latter's accounts of Rhodes and Belgrade as well in his *Süleymanname*.

[75] This decision clearly upset Ferdinand and the Lutherans who assumed the Holy Roman Emperor was no longer concerned with halting the advancing Ottomans. Charles V was in fact worried that the Ottoman armies would target the Italian Peninsula next. Harald Kleinschmidt, *Charles V: The World Emperor* (Phoenix Mill, UK: Sutton Publishing, 2004), 113–14.

[76] Necipoğlu, "Süleyman the Magnificent and the Representation of Power," 401–27; Ágoston, "Information, Ideology, and Limits of Imperial Power," 101.

Barbarossa (d. 1546) *kaptan-ı derya* (admiral) of the Ottoman fleet and *beğlerbeği* (governor general) of *Cezayir-i Bahr-ı Sefid* (the Mediterranean islands) in 1534.[77] This marked a turning point in the Ottoman-Habsburg conflict in the Mediterranean. From this date on, the two empires competed more and more for control of the North African coast. After reestablishing Ottoman control over Naupactus and Coron, Hayreddin Barbarossa successfully seized Tunis and Messina in 1534.

In 1535, an alarmed Charles V personally led a campaign to reclaim Tunis, a port town and the gateway to North Africa. The allied Christian fleet captured La Goletta and Tunis, forcing Hayreddin Barbarossa to abandon his positions in the region.[78] Charles V's conquest of Tunis in 1535 played an important role in the Habsburgs' imperial propaganda by enabling Charles to strengthen and validate his claim to be the Roman Emperor. The conquest marked the beginning of the southward expansion of the empire to Africa to revive the ancient Roman Empire.[79] Even though the city was captured as a result of a local Christian uprising against Hayreddin Barbarossa, Charles's advisers attributed the victory to divine grace. Contemporary artists painted Charles's imperial victory processions through Messina, Rome, and Florence.[80] In these ceremonies, Charles V—not Süleyman—appeared as a world conqueror and the victor of a crusade. Charles had often been depicted as a conqueror who would eventually take Constantinople back from the Turks. But with the success and immediacy of the Tunis campaign, the images of Tunis gradually replaced those of Constantinople, while the Habsburg expansion to Africa was articulated as the revival of the ancient Roman Empire. In contemporary songs and pictures, Charles appeared in the guise of Roman heroes and emperors like Julius Caesar and Alexander the Great.[81] Ottoman accounts, not surprisingly, either completely ignored the loss of Tunis or treated the event as an insignificant setback.[82]

---

77 Idris Bostan, "Cezayir-i Bahr-ı Sefid Eyaleti'nin Kuruluşu (1534)," *Tarih Dergisi* 38 (2002–03): 61–2; 68–75. Imber, *The Ottoman Empire*, 51–3.

78 The Spanish remained in control of Tunis until 1569 when the Ottomans took over. Tunis changed hands several times until 1574 when the Ottomans captured the city. On the naval struggle for Tunis, see J. F. Guilmartin, Jr., *Gunpowder and Galleys: Changing Technology and Mediterranean Warfare at Sea in the Sixteenth Century* (Cambridge: Cambridge University Press, 1974; London: Marine, 2003), 42–56. Citations are to the Marine edition.

79 On Charles V's imperial processions after the Tunis campaign, see Wim Blockmans, *Emperor Charles V, 1500–1558*, trans. Isola van den Hoven-Vardon (London: Arnold Publishers, 2002), 173; Kleinschmidt, *Charles V*, 162–4.

80 Blockmans, *Emperor Charles V*, 173.

81 Ibid; Kleinschmidt, *Charles V*, 162–4.

82 Writing in the 1540s, Matrakçı Nasuh describes the event in his *Süleymanname* as a temporary setback in the Mediterranean. Matrakçı states that Hayreddin and his forces did everything they could and fought in the name of Islam, but were betrayed by the natives of the city. Matrakçı Nasuh, *Süleymanname*, 280b–82a. Barbarossa's biography written by Seyyid Muradi also blames the disloyal natives and the beg of the city who allied with Charles V.

The defeat in Tunis did not deter the Ottoman sultan. While Charles was in Tunis, Süleyman was in Baghdad poised to confront the Safavid state. The Ottoman state renewed efforts to counterbalance the Safavid shahs' assertions regarding sovereignty over Shi'a Ottoman subjects. [83] In 1533, after a treaty with the Habsburgs ended the campaigns in Hungary, Ottoman armies, under the command of İbrahim Pasha, began a long campaign in the East. This was the first Ottoman campaign against the Safavids since Selim's 1514 victory. Süleyman's first imperial campaign in the region, the *sefer-i 'Iraqeyn* (campaign of the two Iraqs), took place between 1533 and 1536. The victorious Süleyman entered Tabriz in 1534 and Baghdad in 1535. The conflict in the East, however, continued intermittently until 1555, and as in Hungary ended without significant economic and territorial gains.[84] Despite the lack of material or territorial acquisitions, these campaigns were clear assertions of universal rule by Süleyman. Charles might have temporarily taken control of Tunis, but Süleyman was the sole ruler of the eastern Mediterranean basin.

Military campaigns in the East and Southeastern Europe did not bring a major victory after the 1540s. Nonetheless, the Ottomans still reigned supreme in the Mediterranean. If Süleyman was to be considered a true successor to the Roman caesars, he had to control the *Mare Nostrum*. Hayreddin Barbarossa had erased the bad memories of the loss of Tunis with a victory over the Venetians in Preveze in 1539. This victory brought important ports such as Kotor on the Dalmatian

---

Muradi continues his story with a description of Hayreddin's campaigns to Mallorca and his safe return to Istanbul. Seyyid Muradi, *Gazavat-i Hayreddin Pasa: Barbaros Hayreddin Pasa'nin Hatiralari*. On Muradi's *Gazavatname*, its different manuscript editions, and its importance as a historical source for the career and campaigns of Hayreddin Barbarossa, see Hüseyin Yurdaydın, "Muradi ve Eserleri," *Belleten* 27 (1963): 453–66; Rhoads Murphy, "Seyyid Muradi's Prose Biography of Hizir Ibn Yakub, alias Hayreddin Barbarossa," *Acta Orientalia Academiae Scientiarum* 54, no. 4 (2001): 519–32; Christine Isom-Verhaaren, "'Barbarossa and His Army Who Came to Succor All of Us': Ottoman and French Views of Their Joint Campaign of 1543–1544," *French Historical Studies* 30, no. 3 (2007): 395–425.

[83] As in the case of Selim I, Süleyman was also obliged to find a legitimate reason to launch this campaign against another Muslim state. Contemporary chroniclers suggest that Safavid Shah Tahmasb's *zandaka* (atheism) and *ilhad* (heresy) peaked during Süleyman's Hungary campaigns. Safavid destruction in the name of Shi'ism had to be ended. Süleyman I's claim to be a champion of Islam necessitated a campaign against the Safavids. Ramazanzade Nişancı Mehmed Pasha, *Tarih-i Nişancı Mehmed Paşa*, 228; Matrakçı Nasuh, *Beyan-ı Menazil-i Sefer-i Iraqeyn-i Sultan Süleyman Han*, ed. Hüseyin G. Yurdaydin (Ankara: TTK, 1976), 2b. For a discussion on the terms *zandaka* and *ilhad* in the Ottoman context, see Ahmet Yaşar Ocak, *Osmanlı Toplumunda Zındıklar ve Mulhidler (15.-17. Yüzyıllar)* (Istanbul: Tarih Vakfi Yurt Yayinlari, 1998).

[84] In 1555, the Amasya treaty left Baghdad to the Ottomans and Tabriz to the Safavids. On the nature of Ottoman-Safavid relations during the reign of Süleyman I, see Adel Allouche, *The Origins and Development of the Ottoman-Safavid Conflict (906–962/1500–1555)* (Berlin: K. Schwarz, 1983), 130–45; M. Fahrettin Kırzıoğlu, *Osmanlılar'ın Kafkas-Ellerini Fethi (1451–1590)* (Ankara: Sevinç Matbaası, 1976).

coast and Monemvasia and Navplion in the Peloponnese under Ottoman control, together with several islands in the Aegean Sea. Following the death of Barbarossa in 1546, Sokollu Mehmed Pasha (d. 1579), a product of the *devşirme* system, became the captain of the Ottoman fleet. Under the command of Sokullu and his two successors, Sinan Pasha and Piyale Pasha, an increasing number of former corsairs—such as Turgut Reis, Uluç Ali Reis, Hasan Reis, and Salih Reis—joined the Ottoman fleet. They continued their raids around North Africa and the Italian Peninsula, only now with the Ottoman sultan's support and blessing. Turgut Reis captured Tripoli in 1551, and Salih Reis, who was appointed governor general of Algiers in 1552, conquered the Spanish stronghold at Bougie. During the summer of 1558, Piyale Pasha led his fleet of 150 vessels from Istanbul to Menorca and destroyed Ciutadella, the main city of the island, located approximately 150 miles from the coast of Spain.[85]

The Mediterranean again became an arena of skirmishes in the 1560s. In 1560, the Ottoman navy (with Piyale Pasha) defeated Andrea Doria at Djerba.[86] With the Ottoman conquest of virtually the entire central and eastern Mediterranean, the realization of imperial visions seemed to be within reach. In 1565, the Ottoman navy laid siege to the island of Malta, which was of strategic importance for further Ottoman expansion into the western Mediterranean, especially around the Spanish coast. Angered by the constant attacks by the Knights of St. John along the trade and pilgrimage routes in the eastern Mediterranean, the Ottoman court sent a large number of troops for the campaign.[87] Food and ammunition was shipped from various locations in Anatolia and from the port cities of Thessalonica, Methoni, and Koroni. Despite the complex effort, the siege ended badly for the Ottomans. After suffering a large number of casualties and failing to penetrate Malta's fortifications, the Ottoman forces eventually withdrew.[88]

The Malta campaign found expression in European circles through three separate channels. European visual, musical, and verbal depictions illustrated the event as a victory of Christianity in Southern Europe over the Muslim Empire.[89] For Ottoman chroniclers, however, the event was merely a defeat, just like the one in Tunis. The sultan himself did not join the campaign; thus, Ottoman chronicles either completely overlooked the event or mentioned only that Ottoman soldiers conquered a castle but failed to capture the island because they were

[85] Katip Çelebi, *Tuhfetü'l-Kibar fi Efsari'l-Bihar*. TKS. Revan Köşkü 1192; Süleymaniye Library, Esad Efendi 2207, 33a–33b.

[86] Andrew C. Hess, "The Moriscos: An Ottoman Fifth Column in Sixteenth-Century Spain," *American Historical Review* 74 (1968): 11.

[87] The Knights of St. John were the inhabitants of Rhodes; they moved to Malta after the fall of Rhodes to the Ottomans in 1522. See especially *6 Numaralı Mühimme Defteri: 972 / 1564–65*, ed. Hacı Osman Yıldırım (Ankara: Başbakanlık Devlet Arşivleri Genel Müdürlüğü Osmanlı Arşivi Daire Başkanlığı, 1995), entry 565.

[88] Arnold Cassola, *The 1565 Ottoman Malta Campaign Register* (Malta: Publishers Enterprises Group, 1994), 325–55.

[89] Ibid., 83–9.

cowardly and fearful.[90] The Malta incident was the last major naval battle in the Mediterranean during the reign of Süleyman I.

In the southeastern corner of the empire, Sultan Süleyman and his ruling elites also tried to take advantage of the legacy that Selim I had left them. During the reign of Süleyman, the Ottomans sought to increase pressure on the Safavids by attempting to secure a foothold in the Red Sea, which would serve as an *entrepôt* into the Indian Ocean. When Süleyman ascended to the Ottoman throne, the Portuguese had already established commercial and political networks in the Indian Ocean.[91] In 1525, a report reached Topkapı Palace warning İbrahim Pasha about Portuguese control over the spice trade. At the beginning of the 1530s, the Ottomans built war galleys in the port city of Suez.[92] To enhance access to the Red Sea, the Ottoman state also discussed the possibility of opening a canal between the Nile and the Red Sea.[93] The canal never materialized, leaving Ottoman naval captains struggling intermittently against the Portuguese in the Red Sea over the next decades.

In the 1530s, convinced by the Ottoman governor of Egypt, Hadım Süleyman Pasha's (d. 1544) pleas, the Ottoman court made several attempts to establish a powerful presence along the southern coast of the Arabian Peninsula and sought a political alliance with Gujarati Muslims.[94] In 1531, an Ottoman fleet helped Gujarat to defend Diu against the Portuguese.[95] This alliance, however, was short-lived. Later, in 1535, Gujarati ruler Sultan Bahadur Shah ceded to the Portuguese the right to construct a fortress in Diu in return for their help against the Mughals.[96] In 1538,

90 Mustafa Selaniki Efendi, *Tarih-i Selaniki*, ed. Mehmet İpşirli (Ankara: TTK, 1999), vol. 1, 10–11.

91 Salih Özbaran, "The Ottomans in Confrontation with the Portuguese in the Red Sea after the Conquest of Egypt in 1517," in *The Ottoman Response to European Expansion*, 94–7.

92 In a report addressing İbrahim Pasha in 1525, Ottoman naval captain Selman Reis warns the grand vizier about Portuguese control over the spice trade in the Indian Ocean, and the threat they pose to the holy cities of Mecca and Medina and even Egypt. Selman Reis urges the grand vizier to take measures as soon as possible. On Selman Reis and his report and its influence on İbrahim Pasha's policies, see Fevzi Ḳurtoğlu, "Amiral Selman Reis Layihasi," *Deniz Mecmuasi* 47, no. 335 (1935): 67–73; Salih Özbaran, "A Turkish Report in the Red Sea and the Portuguese in the Indian Ocean (1525)," *Arabian Studies* 4 (1978): 81–8; Casale, *The Ottoman Age of Exploration*, 36–40.

93 Casale, *The Ottoman Age of Exploration*, 48–9.

94 On Hadım Süleyman Pasha's politics and his infuence at the Ottoman court, see Casale, *The Ottoman Age of Exploration*, 52–83. On Hadım Süleyman Pasha see, Cengiz Orhonlu, "Hadım Süleyman Pasha," *EI*^{2}, last accessed via Brill Online. University of Pittsburgh. 7 November 2011.

95 Casale maintains that this was the first "transoceanic" victory in Ottoman history. Casale, *The Ottoman Age of Exploration*, 48.

96 Sanjay Subrahmanyam, "A Note on the Rise of Surat in the Sixteenth Century," *JESHO* 43, no. 1 (2000): 25.

the Ottoman fleet set sail again to help Gujarat to reconquer Diu. Süleyman Pasha, in fact, led the Ottoman fleet in its excursion into the Indian Ocean and captured Aden in 1538. Despite the size of the Ottoman fleet and the Ottoman admiral's strategic plans, the siege failed thanks in part to poor cooperation between the Gujarati elites and the Ottoman captain.[97] The Gujarati elites, concerned about the demands made by the Ottomans after the victory, showed very little respect to members of the Ottoman fleet. As the fleet left in a rage, the Ottoman admiral complained about the poor quality of Islam in Gujarat.[98]

Hadım Süleyman Pasha, who also served as a grand vizier to Sultan Süleyman between 1541 and 1544, gradually lost his influence at the Ottoman court. In 1544, he was removed from his post due to allegations of financial misconduct during his tenure as the governor of Egypt. In this period, the Ottomans retained their cautious approach in the Indian Ocean. They built naval bases and custom houses in Aden and Mocha, while the Ottoman corsairs and captains continued to challenge the Portuguese in the Indian Ocean between 1538 and 1546.[99] The Ottoman state was able to extend its control over Ta'izz in 1546, and Sana'a in 1547. In 1565, the Ottoman administration divided Yemen into two provinces, Yemen and Sana'a, to control this border zone more effectively. After the division, however, the Ottomans' authority in the area weakened. In 1567, unable to suppress the local Zaydi rebellion, the Ottomans lost Sana'a, Ta'izz, and Aden to Zaydi Imam Mutahhar. While the Ottoman fleet was able to recapture these cities in 1569, it failed to establish a fully functional administrative apparatus in the region.[100]

By the end of the sixteenth century, the Ottomans' presence in the Indian Ocean was negligible.[101] Tardiness in adopting the navigation technology required to sail safely in the Indian Ocean, monetary problems plaguing Constantinople, and poor strategic planning worsened the situation.[102] In 1552, the Grand Vizier Rüstem Pasha commissioned Piri Reis, then the admiral of the Indian Ocean fleet, to conquer the island of Hormuz, which had been a Portuguese base for shipping Persian silk to its ports in India since 1543.[103] Unable to conquer Hormuz and Bahrain, Piri Reis brought his fleet to Basra. Still feeling insecure in these

---

97 Casale, *The Ottoman Age of Exploration*, 59–65.

98 Michael N. Pearson, *Merchants and Rulers in Gujarat: The Response to the Portuguese in the Sixteenth Century* (Berkeley and Los Angeles: University of California Press, 1976), 78.

99 Casale, *The Ottoman Age of Exploration*, 64–80.

100 Hulusi Yavuz, *Kabe ve Haremeyn* İçin *Yemen'de Osmanli Hakimiyeti* (Istanbul: Serbest Matbaası, 1984), 74–99.

101 Salih Özbaran, "Expansion in the Southern Seas," in *Süleyman the Second and His Time*, ed. Halil İnalcık and Cemal Kafadar (Istanbul: ISIS Press, 1993), 241–7.

102 Özbaran, "The Ottoman Empire and the Spice Routes in the Sixteenth Century," in *The Ottoman Response to European Expansion*, 70–72.

103 Newman, *Safavid Iran*, 61.

waters, the admiral abandoned his fleet and headed for Suez with three of his fastest ships. Upon his arrival in Suez, Piri Reis was captured and then executed in Cairo by Ottoman officials.[104] After these incidents, Ottoman attempts to establish a foothold in the Indian Ocean slowed and eventually ceased.

Ottoman Canonical Geography

Süleyman's campaigns during the first half of his reign, particularly those in Hungary and the Mediterranean, spurred the production of geographic and cartographic works in the Ottoman Empire. During this period, Ottoman geographers created an independent body of geographical literature that was accessible to palace dignitaries and the sultan. The available evidence suggests that each geographer acted of his own accord and tried to use his work to earn favors and patronage. At the same time, the court also encouraged geographers to produce maps, campaign diaries, histories, and geographic treatises that organized the available knowledge. In return, these works provided symbolic and geographic support for Ottoman claims to universal sovereignty and categorized the world in accordance with the Ottoman imperial vision. In addition to the works on the Mediterranean, geographical accounts that focused on the capital city and highlighted its centrality in the universe circulated in Constantinople at this time.

An early example of this enterprise during the reign of Süleyman I was Mustafa b. Ali al-Muvakkit's (d. 1571) *İ'lamu'l-'İbad fi A'lami'l-Bilad* (*Public Instructions for the Distances of Countries*).[105] Mustafa b. Ali was a timekeeper (hence his epithet Muvakkit) and served in various imperial mosques around the empire. He received a traditional Ottoman *madrasa* education that involved Qur'anic studies, Islamic law, and jurisprudence as well as Persian and Arabic. As a timekeeper, he was also trained in cosmography and astronomy.[106] He wrote

104 Cengiz Orhonlu, "Hint Kaptanlığı ve Piri Reis," *Belleten* 34, no. 134 (1967): 235–54.

105 There are seven manuscript copies of the work from the sixteenth century. The earliest copy is located in Istanbul Süleymaniye Library: Mustafa b. Ali al-Muvakkit, *İ'lamu'l-'İbad fi A'lami'l-Bilad*, Süleymaniye Library, Hacı Mahmud 5633. For the other sixteenth-century and later manuscript copies, see İhsanoğlu, *Osmanlı Coğrafya Literatürü Tarihi*, vol. 1, 49–51.

106 On Mustafa b. Ali al-Muvakkit and his *İ'lamu'l-'İbad*, see Bursalı Mehmed Tahir, *Osmanlı Mü'ellifleri* (Istanbul: Matba'a-i Amire, 1915–25), vol. 3, 300; Bağdadlı İsmail Pasha, *İzah al-Maknun* (Istanbul, 1945–47), vol. 1, 103; Adıvar, *Osmanlı Türklerinde İlim*, 92–3; İhsanoğlu, *Osmanlı Coğrafya Literatürü Tarihi*, vol. 1, 49–52; Fuat Sezgin, *Geschichte des Arabischen Schrifttums, Band X, Mathematische Geographie und Kartographie im Islam und ihr Fortleben im Abendland* (Frankfurt am Main: Institut für Geschichte der Arabisch-Islamischen Wissenschaften an der Johann Wolfgang Goethe-Universität, 2000), 185–91; Yavuz Unat, "Mustafa İbn Ali el-Muvakkit ve İ'lam el-'İbad fi A'lam el-Bilad (Şehirler Aleminde Mesafelerin Bildirimi) Adli Risalesi," *EJOS* 7, no. 10 (2004): 1–47.

Public Instructions in 1525 and dedicated it to Sultan Süleyman. The text is a coordinate table for 100 cities in the Northern Hemisphere. In addition to the coordinates of these cities, *Public Instructions* provides the distances of 99 cities from Constantinople. Of these 99 cities, those around the eastern Mediterranean, including the geographical regions of Anatolia, the Balkans, the Aegean Sea, the Red Sea. and Egypt, constitute the largest part of this work. In the introduction, the author puts forward his aim: "I have chosen one hundred cities located in the seven climes; some of these are well known to the people of this land and some are not. I also state their distance from Constantinople, which is considered here as the center of all countries."[107]

Although Mustafa b. Ali used a model commonly known to the Muslim world for centuries, he introduced a new center for the world. In the pre-Ottoman Islamic world, Muslim geographers, who were specialized in mathematical geography, prepared coordinate tables to verify and correct Ptolemy's findings. Early examples of such coordinate tables from the eleventh century take Baghdad, the capital city of the Abbasid caliphate and home of the *Bayt al-Hikma* (School of Wisdom), as the central meridian.[108] Written on the eve of Süleyman's successful campaign to Hungary, *Public Instructions* replaces this important political and intellectual center with Constantinople and, in doing so, provides geographic evidence for the Ottoman capital's centrality in the world.[109]

Mustafa b. Ali was not alone in trying to provide geographical scholarship to represent the centrality of the Ottoman Empire and its capital during this period. Four years earlier, in 1521, Piri Reis had completed the first edition of his navigational guide, *Kitab-ı Bahriye* (*Book of Sea Lore*). Later, in response to a special request from İbrahim Pasha, Piri Reis embellished this first version with 214 colorful *portolan* charts and a long introduction in verse. He presented the second version to Süleyman I in 1526, just as the sultan returned from his successful campaign in Hungary.[110] Dedicated and presented to the Ottoman sultan, this first Ottoman Turkish *isolario* offers a more detailed and up-to-date account of the Mediterranean than the similar works prepared by Piri Reis's colleagues in Europe

107 Mustafa b. Ali al-Muvakkit, *İ'lamu'l-'İbad fi A'lami'l-Bilad*, 2a–2b.

108 Fuat Sezgin, *Geschichte des Arabischen Schrifttums, Band X*, 150–73.

109 For further historical analysis of *Public Instructions* and its significance in Ottoman geographical knowledge and imperial claims, see Chapter 2.

110 Piri Reis, *Kitab-ı Bahriye*, Süleymaniye Library, Ayasofya 2612. This copy in the Süleymaniye Library is attested as the most complete copy. It was donated to the library by Mahmud I (r. 1730–54). Today there are 45 manuscript copies of this second version in libraries around the world. For the list of these locations, see Mine Esiner Özen, *Piri Reis and His Charts* (Istanbul: N. Refioğlu, 1998), 17, 20–22. There are 5,658 manuscript maps in the 43 known copies of the work. Nine of these copies are located in libraries in England, Italy, France, Germany, and Australia. The rest are listed in the catalogs of different libraries and museums in Istanbul. Among the 45 manuscripts of this work, two include only text, while three consist exclusively of maps.

during the same period.[111] In the *Book of Sea Lore*, the sea captain drew on his more than 30 years of experience as a seafarer. He describes his own observations and experiences as a sailor, makes practical sailing suggestions, and describes the landmarks and facilities offered in each port. His narrative or technical passages are usually complemented by short write-ups about the local history and mythology.

The works of both Piri Reis and Mustafa b. Ali lent support to Süleyman's claims to be the new world conqueror and Holy Roman Emperor. As Süleyman's ruling elites promoted these claims, Piri Reis, Mustafa b. Ali, and many other geographers assisted this enterprise by offering works that defined the Ottoman Empire as the center of the universe. Through their work, they not only provided the practical knowledge required for Ottoman military expansion but also enhanced the Ottomans' awareness of the expanding world.

More than a decade later, Matrakçı Nasuh (d. 1564), a mathematician and geographer, compiled an account of Süleyman's first imperial campaign in the East entitled *Mecmu'-i Menazil: Beyan-ı Sefer-i Iraqeyn-i Sultan Süleyman Han* (*The Collected Stages: The Description of Sultan Süleyman's Campaign in the Two Iraqs*).[112] Matrakçı Nasuh, a protégé of grand vizier Rüstem Pasha, entered the palace service as a *devşirme* and was a product of Ottoman schooling and administration. In 1520, after Matrakçı made a name for himself in Constantinople as a talented miniaturist, mathematician, and sword-master, Süleyman commissioned him to translate Arab historian Muhammad Ibn Jarir al-Tabari's tenth-century universal history, *Tarikh al-Tabari*, into Ottoman Turkish. Matrakçı Nasuh translated the work, made emendations, and added a section on the downfall of the Abbasids, and on the origin of the Turks, Ghaznavids, Seljuks, and Ottomans.[113] Simultaneously,

---

111 George Tolias, "Isolarii, Fifteenth to Seventeenth Century," in *The History of Cartography, Vol. 3, Bk. 1, Cartography in the European Renaissance*, ed. Harley and Woodward, 269–70. For a detailed study of the *Book of Sea Lore*, see Chapter 3.

112 On Matrakçı Nasuh's life and career, see J. M. Rogers, "Itineraries and Town Views in Ottoman Histories," in *The History of Cartography, Vol. 2, Bk. 1, Cartography in the Traditional Islamic and South Asian Societies*, ed. Harley and Woodward, 228–55; Hüseyin Yurdaydın, "Introduction," in Matrakçı Nasuh, *Beyan-ı Menazil-i Sefer-i Irakeyn-i Sultan Süleyman Han*, ed. Yurdaydın; and Hüseyin Yurdaydın, *Matrakçı Nasuh* (Ankara: Ankara University Press, 1963). On different interpretations of the *Mecmu'-i Menazil*, see Sadi Selen, "16. Asırda Yapılmış Anadolu Atlası," in *II. Türk Tarih Kongresi, Istanbul, 20–25 Eylül 1937, Kongreye Sunulan Bildiriler* (Istanbul: Kenan Matbaasi, 1943); Franz Taeschner, "The Itinerary of the First Persian Campaign of Sultan Süleyman, 1534–36, According to Nasuh al-Matraki," *Imago Mundi* 13 (1956): 53–5; Norman J. Johnston, "The Urban World of the Matraki Manuscript," *Near Eastern Studies Journal* 30, no. 3 (1971): 159–66; İffet Orbay, "Istanbul Viewed: The Representation of the City in Ottoman Maps of the Sixteenth and Seventeenth Centuries" (PhD diss., MIT, 2001), 52–70. Kathryn A. Ebel, "Representations of the Frontier in Ottoman Town Views of the Sixteenth Century," *Imago Mundi* 60, no. 1 (2008): 1–22.

113 Hüseyin Yurdaydın, "Introduction," in Matrakçı Nasuh, *Beyan-ı Menazil-i Sefer-i Irakeyn-i Sultan Süleyman Han*, 119–25.

he embarked upon a project of writing the history of Süleyman. Later on, Matrakçı copied the *Beyan-ı Menazil* into this later work, entitled *Süleyman-name*.

At first glance, the *Beyan-ı Menazil* is not very different from the previous campaign diaries written for and circulated in Constantinople. It simply describes the campaign and the major stops en route from Constantinople to Baghdad. Matrakçı Nasuh completed his work in 1537 and presented it to Süleyman I.[114] What is striking about the work is that Matrakçı Nasuh composed the *Beyan-ı Menazil* during the actual campaign and later illustrated it with 130 miniature depictions of the major stations.[115] Most of the miniature paintings depict the Anatolian urban centers that had been part of the Ottoman administration since the fifteenth century. In addition, Matrakçı Nasuh provided illustrations of Aleppo, Basra, Baghdad, and Tabriz, new additions to the Ottoman territories. The text that accompanies these depictions gives a lengthy description of the cities and their topography, public facilities, and holy places. It also highlights the role of the sultan in this campaign and portrays him as the dispenser of justice throughout his domains and beyond. Unlike his predecessors, therefore, Matrakçı Nasuh's work—which was in essence a pictorial account of Süleyman's campaign and the empire's latest conquest—conveyed to the Ottoman sultan not only the breadth of Ottoman territories, but also what many of these lands were like.[116]

Written during the imperial campaign against the Safavids, the *Beyan-ı Menazil* is a remarkable contribution to Ottoman geography. It distinguishes itself from the other geographical works with its amount and quality of information, which was crucial for a geographical and imperial understanding of the extent of the Ottoman domains. The work demonstrates clearly the ways in which Ottoman geographers dealt with territorial expansion, integrated newly acquired domains into the empire, and presented their findings to the Ottoman sultan. In doing so, it delineates the relationship between the imperial vision and geographical knowledge during Süleyman's reign. It is not a coincidence that the work opens with a depiction of Constantinople, which the author acknowledges as the center of the empire and of the world.

Within a decade, Sultan Süleyman had at his disposal at least three works that mapped and described the lands of the expanding Ottoman Empire and that placed Constantinople at the center of the universe. These three works were written for

---

[114] Only one manuscript copy of this work has survived. Matrakçı Nasuh, *Beyan-ı Menazil-i Sefer-i 'Iraqeyn-i Sultan Süleyman Han*, Istanbul University Library, TY. 5964. The facsimile edition of the work was published by Hüseyin Yurdaydın.

[115] Matrakçı prepared similar chronicles, also illustrated with miniature depictions: Matrakçı Nasuh, *Fethname-i Karabuğdan* (1538), TKS. Revan Köşkü 1284/2; idem, *Tarih-i Feth-i Sikloş ve Estergon ve İstun- Belgrad* (1543), TKS. Hazine 1608; idem, *Tarih-i Sultan Bayezid* (1540–50), TKS. Revan Köşkü 1272. While scholars have not yet been able to agree on the identity of the miniaturists, it was Matrakçı Nasuh who wrote the text of *Beyan-ı Menazil* and also painted its miniatures.

[116] For a detailed historical analysis of the work and its importance for the Ottoman geographical knowledge, see Chapter 2.

the sultan when the Ottoman court, under the guidance of İbrahim Pasha, began refashioning the physical appearance of Constantinople, as well as the imperial palace and its ceremonies.[117] Under his supervision, the Topkapı Palace and imperial ceremonies were redesigned to highlight the Ottomans' military victories and, more importantly, the magnificent persona of the sultan and the universality of his empire.[118] Through large-scale restoration activities, most buildings in the second courtyard of the imperial palace—home to the treasury, the tower of justice, and the council hall—as well as the chamber of petitions of the third courtyard, were either rebuilt or remodeled.[119] The Hippodrome, which stood in the center of the city during Byzantine times, reopened for public spectacles and became an integral part of imperial wedding and circumcision ceremonies.[120] İbrahim Pasha, who was married to Süleyman's sister Hadice, even built his own palace right next to the Hippodrome.[121]

The Ottoman geographers of this period, namely Piri Reis, Mustafa b. Ali, and Matrakçı Nasuh, consolidated and rendered visible the Ottomans' evolving geographical consciousness. In the process, they created canonical works that recast the available knowledge for their audience, which consisted of the Ottoman intellectuals, ruling elites, and, importantly, the sultan. In this way, geographers supported İbrahim Pasha's agenda to portray the Ottoman Empire as a world empire with a strong imperial vision for universal sovereignty. All three geographers prepared their works for the Ottoman sultan. Piri Reis was a protégé first of İbrahim Pasha and later Rüstem Pasha, who also patronized Matrakçı Nasuh. These geographers were politically and intellectually active in the same circles. It is highly plausible that they also knew each other and each other's works very well. It is not surprising that in their works they promoted the prominent imperial vision of the time which prioritized Constantinople as the center of the world and the Mediterranean as the main focus of Ottoman political, economic, and ideological concerns. The imperial culture at the Ottoman court was, however, bound to change in the next decades.

---

[117] Necipoğlu, *Architecture, Ceremonial, and Power*, 22–3.

[118] Ibid., 22–30.

[119] Ibid., 76–110.

[120] Turan, "The Sultan's Favorite," 177–8; Çiğdem Kafescioğlu, *Constantinopolis/Istanbul: Cultural Encounter, Imperial Vision, and the Construction of the Ottoman Capital* (University Park: Penn State University Press, 2009), 53–109.

[121] Süleyman, in fact, contracted marriages between his three sisters Hadice, Şah, and Fatma and his three grand viziers, İbrahim, Lütfi, and Kara Ahmed. Peirce, *The Imperial Harem*, 66–8. On İbrahim Pasha's palace and its significance, see Turan, "The Sultan's Favorite," 145–79.

A Somber Image and a Sober Policy

By the 1540s, it was clear that neither Charles V nor Süleyman I could establish universal rule.[122] The military confrontation in Hungary ended in a stalemate and Charles V withdrew to a monastery in 1556. The aging Süleyman, meanwhile, stopped participating in imperial campaigns and royal processions except for those connected to the Friday prayers. He delegated most of his power to imperial bureaucrats and withdrew to his own quarters in the palace.[123] As the imperial campaigns fizzled on both fronts, an aging Süleyman faced personal setbacks. After a string of military defeats and with rumors of his loss of power flying, the sultan felt compelled to order the execution of his beloved grand vizier, İbrahim Pasha, in 1536.[124] Shaken by this event and later by the death of his favorite son, Mehmed (d. 1543), the sultan began to espouse a "somber image," placing great emphasis on Muslim piety.[125]

The second half of Süleyman's reign was a remarkable episode in the history of the Ottoman Empire due to developments in bureaucracy and administration. The period, reminiscent of the first half of Süleyman's reign, saw the rise of another grand vizier who shaped imperial culture and policy in the Ottoman court. Rüstem Pasha, who was raised in the *devşirme* system, was quickly promoted through the ranks thanks to the patronage of Süleyman's wife Hürrem Sultan. Later married to the sultan's only surviving daughter, Mihrimah Sultan, Rüstem Pasha became the grand vizier in 1544 and held that position intermittently until his death in 1561.[126] The formation of an institutionalized bureaucracy was an integral part of Rüstem's political agenda during this period. While "a small and undifferentiated body of scribes" had dealt with the bureaucratic needs of the empire during the reigns of Bayezid II and Selim I, by the end of the reign of Süleyman I, the financial branch of the bureaucracy under the *başdefterdar* (chief treasurer) had grown in size, and the chancery office of the *kalemiye* (bureaucracy) had been consolidated. As the offices of the bureaucracy gained an institutional framework, promotion patterns were standardized, and the central administration was expanded. During the same period, the Ottoman state apparatus also codified dynastic law and reconciled it with Islamic Law. The institutionalization and consolidation of the bureaucracy, which saw to all functions of government,

---

122 Kleinschmidt, *Charles V*, 114.

123 Cornell H. Fleischer, "Preliminaries to the Study of the Ottoman Bureaucracy," *Journal of Turkish Studies* 10 (1986): 135–42. Necipoğlu, *Architecture, Ceremonial, and Power*, 29–30.

124 Peirce, *The Imperial Harem*, 74–5.

125 Ibid., 76.

126 Rüstem Pasha was forced to retire in 1553 due to his role in the execution of Süleyman's son Mustafa. Rüstem Pasha returned to the post in 1555. Christine Woodhead, "Rüstem Pasha," *EI*[2], last accessed via Brill Online. University of Pittsburgh. 7 November 2011. See also, Casale, *Ottoman Age of Exploration*, 84–5. On the importance of royal marriages in the Ottoman court, see Peirce, *The Imperial Harem*, 65–72.

allowed Süleyman I to withdraw ever more deeply into seclusion. With the sultan no longer playing an active role in governance, the Ottoman bureaucratic establishment employed the *kanun* (imperial law) on Süleyman's behalf as a sign of his rule and unchallenged authority.[127]

This period of sober and pious imperial tradition also saw the creation of a unified Ottoman style in architecture, which expressed both the orthodox Islamic character of the empire and the somber persona of the sultan.[128] In contrast to the eclectic vision of the previous decades, buildings constructed from the 1540s onward emphasized Ottoman colors and motives.[129] One of the foremost patrons of this period, the grand vizier Rüstem Pasha commissioned several mosques and *madrasas* throughout the empire and also encouraged the Ottoman manufacture of textiles and ceramic tiles. During his tenure, and in contrast to the policies of his predecessor, İbrahim Pasha, he restricted the import of luxury goods from Europe. Thus, the local manufacturers of textiles in Bursa and of ceramic tiles in İznik and Kütahya provided the decoration materials for the newly erected buildings in Istanbul. The best example of this development is the Rüstem Pasha Mosque in Istanbul. The mosque's interior decoration—replete with tiles from İznik—reflects the changes in the Ottomans' aesthetic vision.

During this period, the *ser-mimar* (chief architect) Sinan (d. 1588) oversaw numerous construction projects sponsored by Süleyman I and his ruling elites in Istanbul.[130] From 1539—when he became the chief architect—until his death in 1588, Sinan designed around 200 buildings in the city including mosques, *madrasas*, baths, hospices, and palaces. These buildings substantially changed the skyline of the city.[131] The mosque complexes in different corners of Istanbul adorned the Ottoman "threshold of felicity" with lofty domes and slender minarets

[127] Cornell Fleischer, *Bureaucrat and Intellectual in the Ottoman Empire: The Historian Mustafa Ali (1541–1600)* (Princeton, NJ: Princeton University Press, 1986), 191–252; idem, "The Lawgiver as Messiah," 166–70. During this period, the Şeyhülislam Ebussu'ud Efendi (1490–1574) and the chancellor Celalzade Mustafa (1490–1567) systematically codified, compiled, and reconciled the *kanun* (imperial law) with Islamic Law. See also Colin Imber, *Ebu's-Su'ud: The Islamic Legal Tradition* (Palo Alto, CA: Stanford University Press, 1997).

[128] Gülru Necipoğlu, "A Kanun for the State, a Canon for the Arts," in *Soliman le Magnifique et son Temps*, ed. Giles Veinstein, (Paris: La Documentation Française, 1992), 198–216.

[129] Ibid.

[130] On Sinan and his career as the chief architect and his achievement in creating a unique Ottoman style in architecture, see Gülru Necipoğlu, *The Age of Sinan: Architectural Culture in the Ottoman Empire* (Princeton, NJ: Princeton University Press, 2005).

[131] The Şehzade Mehmed Mosque (1543–48) and the Süleymaniye (1550–57) in particular shaped the historical peninsula and dominated the Golden Horn and the Marmara Sea. Doğan Kuban, *İstanbul Bir Kent Tarihi: Bizantion, Konstantinopolis, İstanbul*, trans. Zeynep Rona (Istanbul: Türkiye Ekonomik ve Toplumsal Tarih Vakfı, 1996), 237.

reaching to the sky.[132] This panorama not only distinguished the capital city from the provinces of the empire but also articulated the empire's unparalleled magnificence.

Sinan's monuments established the canon for an architecture that was distinct from its European and Safavid counterparts.[133] In constructing these edifices, Sinan enhanced and amended already established patterns. He revised and refined the architectural plans of extant buildings and integrated architectural materials from all over the empire into new structures.[134] In doing so, he drew on aesthetic forms that had been established before his time, and he standardized monuments so that they were now similar throughout the empire.[135] Creating a unified Ottoman art recognizable throughout the empire ensured "readability and repeatability" for its audience.[136] The Topkapı Palace's royal architects—headed by Sinan, who supervised most of the construction projects—further consolidated "the unity of this imperial style."[137] Sinan's works gave architectural form to the growing Ottoman belief that Constantinople was the center of the universe. By the time of Süleyman's death, this idea was reflected in historical and geographical accounts, and in the very nature of the city's design.

Ottoman Discovery of the New Worlds

In the second half of Süleyman's reign, when Sinan and the royal architects of the Palace created a canonical style of Ottoman architecture, Ottoman geographers began to form a canon of Ottoman geographical knowledge. Rüstem Pasha's role in this enterprise was inevitable. Even though the grand vizier did not try to institutionalize geographical production, his patronage gave individual geographers an incentive to produce works for the Ottoman court and to shape the political vision of their audience. While the major theme of the geographical works of this period was still the centrality of Constantinople and the importance of the Mediterranean for imperial claims, the latest discoveries and political

---

[132] Necipoğlu's latest work on Sinan and his mosque complexes offers an exquisite analysis of how the chief architect developed a hierarchical system of architectural representation in his Friday mosques to articulate the status and aspirations of his patrons. These complexes also contributed to the urban structure of the city. See Necipoğlu, *The Age of Sinan*, 189–506.

[133] Kuban, İstanbul Bir Kent Tarihi, 236–7.

[134] For example, three of the four minarets of the Süleymaniye mosque were from Kıztaş, Alexandria, and Balbek. They were all transported to Istanbul by ship. Mustafa Sa'i Çelebi, *Tezkiretü'l-Bünyan*, Süleymaniye Library, Hacı Mahmud Efendi 4911, 9a–9b.

[135] Necipoğlu, "The Süleymaniye Complex in Istanbul: An Interpretation," *Muqarnas* 3 (1985), 112.

[136] Necipoğlu, "Challenging the Past: Sinan and the Competitive Discourse of Early Modern Islamic Architecture," *Muqarnas* 10 (1993), 173.

[137] Necipoğlu, "Süleymaniye Complex in Istanbul," 112.

developments in the Atlantic and Indian oceans were also introduced in geographers' accounts. As Ottoman intellectuals tried to comprehend the scale of geographical discoveries and integrate these new and uncharted territories into their geographical vision, they also continued to promote the political agendas of their patrons.

In 1534, French cosmographer Orance Fine (d. 1555) prepared a heart-shaped woodcut world map.[138] On the map, the latest geographical discoveries in the New World were carved in a conventional shape, and the world was presented as a unified form without political or religious borders. In 1559, a world map based on Fine's 1534 woodcut design was prepared in Ottoman Turkish. Entitled *The Representation of the Whole World Designed in Its Entirety*, the map was made in Venice by a certain Hacı Ahmed of Tunis. Hacı Ahmed was a fictitious character; the compiler of the work was neither an educated Tunisian nor an Ottoman.[139] Unfortunately, no sixteenth-century copies of the map survive. The woodblocks remained in the Venetian archives, and only in 1795 were 24 impressions made.[140] Despite the misspellings and grammatical errors in the text, the language and correct honorifics used for the Ottoman sultan strongly suggest that the map was prepared with the Ottoman market in mind.[141] Recent research in the Venetian archives has shown that Prince Selim, son of Süleyman I and the future Sultan Selim II, ordered world maps from the workshops in Venice in the early 1550s, right before the succession struggle began. We know today that all three of Süleyman I's sons were interested in Venetian cartographical productions as symbols of prestige and power during their struggle for the Ottoman throne.[142] The *Hacı Ahmed Map* and its production story epitomize how geographers sought the patronage of Ottoman princes and sultans for their latest works. Although the geographer in this case was Venetian, he most likely had an Ottoman patron, which

---

138 On the importance of this map for early modern geographical knowledge and the influence of geographical discoveries on the production of this world map, see Giorgio Mangani, "Abraham Ortelius and the Hermetic Meaning of the Cordiform Projection," *Imago Mundi* 50 (1998): 59–82; Denis Cosgrove, *Apollo's Eye: A Cartographic Genealogy of the Earth in the Western Imagination* (Baltimore: Johns Hopkins University Press, 2001), 133–4; Frank Lestringant and Monique Pelletier, "Maps and Descriptions of the World in Sixteenth-Century France," in *The History of Cartography, Vol. 3, Bk. 2, Cartography in the European Renaissance*, ed. Harley and Woodward, 1464–8.

139 V. L. Ménage, "The Map of Hajji Ahmed and Its Makers," *BSOAS* 21 (1958): 299–311.

140 Robert Karrow, *Mapmakers of the Sixteenth Century and Their Maps: Bio-Bibliographies of the Cartographers of Abraham Ortelius, 1570* (Chicago: Published for the Newberry by Speculum Orbis Press, 1993), 172.

141 Ménage, "The Map of Hajji Ahmed"; Jerry Brotton, "Printing the World," in *Books and Sciences in History*, ed. Marina Frasca-Spada and Nick Jardine (Cambridge: Cambridge University Press, 2000), 35–48.

142 Benjamin Arbel, "Maps of the World for Ottoman Princes? Further Evidence and Questions Concerning 'The Mappamondo of Hajji Ahmed,'" *Imago Mundi* 54 (2002): 21–30.

suggests that—politically and symbolically—maps had taken on new meaning. Previously, geographers approached the incumbent sultan to offer their work; now, contenders for the Ottoman throne actively sought out maps to aid in their bid for the empire's highest office. For members of the Ottoman court, having a reputation as "a prestigious patron for maps" prepared in Europe became an important factor in earning the support of one of the factions in Constantinople.[143]

Around the same time that the *Hacı Ahmed Map* was prepared for the Ottoman market, an Ottoman sea captain, Seydi Ali Reis, compiled two accounts on India and the Indian Ocean under the patronage of the grand vizier, Rüstem Pasha.[144] Seydi Ali Reis's repute in Ottoman history partially stems from his success in safely escorting the Ottoman fleet—which Piri Reis had abandoned in Basra—to the port of Surat in India in 1554. He completed his first work entitled *Kitab-ı Muhit* (*Book of the Ocean*) in Ahmedabad the same year. The work is a navigation manual for the Indian Ocean and is dedicated to Süleyman I. Although the work went almost unread in the sixteenth century, it attracted a lot of interest among the orientalists of the nineteenth century.[145] Its 10 chapters cover a wide range of subjects, including techniques and instruments employed to navigate in the

143 For the significance of the *Hacı Ahmed Map* for the Ottoman understanding of the New World, see Chapter 4.

144 According to Casale, Seydi Ali Reis was a prominent member of the faction of the grand vizier Rüstem Pasha, who sought to control the border regions of the Red Sea and the Persian Gulf and to shift military activity from the sea to the hinterland. Casale, *The Ottoman Age of Exploration*, 88.

145 In the nineteenth century, several Turkologists translated sections from the *Book of the Ocean*. Hammer-Purgstall translated the chapters on marine astronomy and physical geography in a series of articles dated between 1834 and 1839: J. von Hammer-Purgstall, "Extracts from the Moh'it, that Is the Ocean, A Turkish Work on Navigation in the Indian Seas," *Journal of the Asiatic Society of Bengal* 3, no. 35 (1834): 545–53; 5, no. 56 (1836): 441–68; 6, no. 70 (1837): 805–12; 7, no. 81 (1838): 767–80. Later, Maximilian Bittner translated the sections on the coasts of India and the famous ports of the Indian Ocean. M. Bittner with W. Tomaschek, "Die topographischen Capitel des indischen Seespiegels Mohit, mit einer Einleitung sowie mit 30 Tafeln versehen," in *Studies on the Works of Sidi Ali*, 129–254. In 1894 and 1895, Luigi Bonelli prepared the Italian editions for the fourth, sixth, and seventh chapters of the work, using the manuscript copies in Vienna and Naples, Luigi Bonelli, "Del Muhit o'Descrizione dei mari delle Indie' dell'ammiraglio turco Sidi 'Ali detto Kiatib-i Rum," *Rendiconti dell'Accademia dei Lincei, Classe di scienze morali, serie quinta* 3 (1894): 751–77; 4 (1895): 36–51. On this edition, see Maximilian Bittner, "Zum 'Indischen Ocean des Seidi 'Ali'. Bemerkungen zu einer Übersetzung aus dem Türkischen," in *Studies on the Works of Sidi Ali*, vol. 16 of *Islamic Geography*, ed. Fuat Sezgin et al. (Frankfurt: IGAIW, 1992), 113–29; Svat Soucek, "Sidi Ali Reis," *EI²*, last accessed via Brill Online. University of Pittsburgh. 7 November 2011. On Seydi Ali Reis and the manuscript copies of the *Book of the Ocean* and its importance as a geographical work, see Adıvar, *Osmanlı Türklerinde İlim*, 77; İhsanoğlu, *Osmanlı Coğrafya Literatürü Tarihi*, vol. 1, 35–8; Svat Soucek, "Sidi Ali Reis," *EI²*; Seydi Ali Reis, *Book of the Indian Ocean: Kitab al-Muhit*, ed. Fuat Sezgin (Frankfurt: IGAIW, 1997); Casale, *The Ottoman Age of Exploration*, 98–102.

Indian Ocean; the history, geography, and anthropology of the region; and new geographical discoveries, all conveyed in an elegant and eloquent language. The work does not contain any maps.

Looking at Seydi Ali Reis's manual, it is clear that he was driven by a desire to demonstrate his professional knowledge in navigation to the Ottoman sultan. The sea captain also takes the opportunity to draw the sultan's attention to the Indian Ocean while at the same time carefully positioning it on the periphery of the Ottoman world. As Rüstem Pasha's protégé, Seydi Ali Reis confirms that the Ottomans had to keep an eye on the Indian Ocean, primarily because of the Portuguese and their activities there. Seydi Ali Reis repetitively cautions his readers about the dangers of sailing in the area and urges them to stay away from these troubled waters.[146]

In 1557, Seydi Ali Reis composed another geographical account, a diary of his journey from Gujarat to Istanbul, which is entitled *Mir'atü'l Memalik* (*Mirror of Lands*).[147] It appears that the author prepared this work for the grand vizier Rüstem Pasha soon after he returned from India. In this diary account, Seydi Ali Reis describes the geography, society, and local rulers of India, and the Portuguese engagements in the Indian Ocean. *Mirror of Lands* has to be viewed and interpreted in the context of a slowing Ottoman expansion and a tapering off of the Ottoman claim for world conquest. In Seydi Ali Reis's account, the reader learns about the Muslim communities in India and the lands through which Seydi Ali Reis passed en route from Gujarat to Istanbul. Written during the sober and pious period when Süleyman I had gone into seclusion, the diary confirms that the sultan's spiritual authority was accepted even in this distant part of the world, and it carefully places India on the periphery of the Ottoman Empire.[148]

---

146 For a detailed analysis of this work and samples of these passages, see Chapter 4.

147 Several European-language translations of the work appeared during the nineteenth century. In 1815, H. F. von Diez translated *Mirror of Lands* into German. H. F. von Diez, *Denkwürdigkeiten von Asien* (Berlin: Nicolai Verlag, 1815), vol. 2, 133–276. Its French translation appeared in *Journal Asiatique* in 1826. In this article, M. Morris partially translated the German text by Diez into French. M. Morris, "Miroir des pays, ou relation des voyages de Sidi Aby fils d'Housain, nommé ordinairement Katibi Roumi, Amiral de Soliman II," *Journal Asiatique* 9 (1826): 27–286. More than 60 years after this translation was completed, A. Vambéry translated the printed copy of the original text into English. Seydi Ali Reis, *The Travels and Adventures of the Turkish Admiral Sidi Ali Reis in India, Afghanistan, Central Asia, and Persia, During the Years 1553–1556*, trans. A. Vambéry (London: Luzac & Co., 1899). After the foundation of the republic, the work was also translated into Modern Turkish. An abridged Turkish translation was prepared by Mustafa Nihat Özön and Hayrullah Örs in 1935. Mustafa Nihat Özön and Hayrullah Örs, *Hindeli'nden İstanbul'a* (Ankara, 1935). Another short and simplified Turkish translation appeared in the second half of the twentieth century. Seydi Ali Reis, *Mir'at ül-Memalik (Ülkelerin Aynası)*, ed. Necdet Akyıldız (Istanbul: Tercüman, 1975). Finally, in 1999, Mehmet Kiremit published a critical edition of the work. Seydi Ali Reis, *Mir'at ül-Memalik: İnceleme, Metin, İndex*, ed. Mehmet Kiremit (Ankara: TTK, 1999).

148 For a detailed analysis of this work and its careful placement of India on the peripheries of the Ottoman Empire, see Chapter 4.

It is clear from Seydi Ali Reis's account that India was only a marginal concern in Constantinople. In the 1560s, when Seydi Ali Reis's works went almost unread, the court was teeming with world maps, navigation charts, and *portolan* atlases. The earliest of these atlases, *Walters Sea Atlas* (ca. 1560), features seven *portolan* charts and a world map based on European cartographic concepts. The atlas was probably prepared in Italy for a readership in the Ottoman Empire. Although we cannot identify the cartographer or the original recipient of the atlas, the elaborate illustrations on the charts suggest a wealthy and prestigious patron, possibly a member of the Ottoman court.[149] Since the atlas entered the Ottoman palace when the struggle against the Spanish Habsburgs was intensifying in the Mediterranean, its charts focus exclusively on this region.[150]

From one perspective, all of the works produced during the reign of Süleyman can be seen as individual efforts by a few intellectuals, sailors, and officials to impress the Ottoman court elites or the sultan. However, a closer look reveals that these individual geographers helped to create and later enhanced the Ottomans' understanding of the world, while at the same time articulating the political claims of their patrons. Maps, charts, travel accounts, and geographical treatises from this period had multiple purposes but one consistent effect: the creation and promotion of a unique Ottoman geographical consciousness, which was readable and comprehensible by the Ottoman intellectuals, the sultan, and his ruling elites. Written during the age of Ottoman territorial expansion, and in many ways inspired by this success, Ottoman geographical works both reflected and shaped the Ottoman understanding of the ever-expanding world and the empire's central place within it. The accounts on the Indian Ocean and the New World that were produced under the patronage of the court elites reiterated that the Indian Ocean and the New World were faraway lands, and although not as important as the Mediterranean, the Ottoman court should learn about them to improve its geographical understanding of the world. It is the contention of this book that given their distance from Constantinople, the Ottoman state did not view these oceans and the activities of the Europeans in them as threats to the empire's imperial aspirations. Instead, they were seen as peripheral regions, a status that some Ottoman geographers reflected.

Closure of the Sixteenth Century: The Ottoman Imperial Image Challenged

In 1566, Süleyman I died behind the walls of the fortress of Szigetvar during his last campaign in Hungary. His grand vizier, Sokullu Mehmed, continued the siege and conquered the town. During the reigns of Süleyman's successors, Selim II (r. 1566–74) and Murad III (r. 1574–95), Ottoman military campaigns

[149] Thomas Goodrich, "The Earliest Ottoman Maritime Atlas: The Walters Deniz Atlasi," *Archivum Ottomanicum* 11 (1986): 25–50.

[150] For a detailed analysis of this atlas and its significance for the Ottoman imperial claims in the Mediterranean, see Chapter 3.

slowed down, as both the Spanish and the Ottomans accepted their allotted share of the Mediterranean in the second half of the sixteenth century. As the Spanish took control of the western Mediterranean, the Ottoman navy policed the eastern Mediterranean. In 1570, the Ottomans laid siege to Cyprus, conquering Nicosia the same year and Famagusta in 1571. Only a few months after this long-awaited conquest, at the Battle of Lepanto, the Holy League destroyed most of the Ottoman fleet. Within a year, under the grand vizier Sokullu Mehmed Pasha's directions, the Ottoman arsenal reconstructed the fleet. In 1574, another naval expedition reconquered Tunis from the Spanish.[151]

In 1574, Murad III ascended the Ottoman throne. The new sultan continued his father Selim II's aggressive policies in the Mediterranean. In North Africa, the Ottomans succeeded in installing an Ottoman client as a ruler in Morocco. This political maneuver gave the Ottomans control over the whole of North Africa until 1578 when Portugal invaded Morocco. In 1580, with a peace treaty, both the Habsburgs and the Ottomans freed themselves from concentrating on the Mediterranean any further. The border between the Spanish Habsburgs and the Ottomans was stabilized.[152] Although the Ottomans were officially still in control of Tripoli, Tunis, and Algiers, it is impossible to argue that North Africa was fully under the control of the Ottoman administration.

The reign of Murad III in particular was a period of ongoing political and financial crisis for the Ottomans as the empire was forced to fight wars on two fronts. The Ottomans were in intermittent conflict with the Safavids in Persia from 1579 to 1639, while in Hungary a long and costly war with the Habsburgs raged between 1593 and 1606.[153] Confronted with financial crisis, the Ottoman state levied new taxes on peasants, borrowed money from internal moneylenders, and debased the value of silver coins. Irate over lost wages, the cavalrymen of the Porte (*kapı*kulu sipahis) revolted in Istanbul.[154] Contemporary historians based in the Ottoman capital interpreted these events as a sign of decline.[155]

---

151 Imber, *The Ottoman Empire*, 63.

152 Andrew Hess argued that this stabilization brought the "unique divergence of the Latin Christian and Turko-Muslim civilizations." Hess, *The Forgotten Frontier: A History of the Sixteenth-Century Ibero-African Frontier* (Chicago: University of Chicago Press, 1987), 207. Baki Tezcan adds that this divergence "was accompanied by an implicit recognition of equality between the coercive powers of the two civilizations." Tezcan, "The 'Frank' in the Ottoman Eye of 1583," in *The Turk and Islam in the Western Eye, 1450–1750*, ed. James G. Harper (Burlington, VT: Ashgate, 2011), 279.

153 Colin Imber, *The Ottoman Empire*, 63–72; idem, "Frozen Legitimacy," in *Legitimizing the Order: The Ottoman Rhetoric of State Power*, ed. Hakan Karateke (Leiden: Brill, 2005), 101–2.

154 Cemal Kafadar, "When Coins Turned into Drops of Dew and Bankers Became Robbers of Shadows: The Boundaries of Ottoman Economic Imagination at the End of the Sixteenth Century" (PhD diss., McGill University, 1986).

155 Cemal Kafadar, "The Myth of the Golden Age: Ottoman Historical Consciousness in the Post-Süleymanic Era," in *Süleyman the Second and His Time*, ed. Halil Inalcık and Cemal Kafadar (Istanbul: Isis Press, 1993), 37–48; Necipoğlu, *The Age of Sinan*, 256–7.

The *realpolitik* in this period prompted action by Ottoman dignitaries. Murad held fast to the claims of his predecessors to universal sovereignty.[156] At the same time the Ottoman court proposed unconventional and creative solutions to address the rising Shi'i power beyond the Ottomans' eastern frontier. There was no doubt that Safavid economic, military, and political achievements under Shah Abbas I (r. 1587–1629) seriously challenged the Ottoman presence and—most importantly—its legitimacy in the East.[157] The Safavid chroniclers in Isfahan articulated the Safavids' supremacy vis-à-vis the two Sunni powers: the Ottomans in Anatolia and Iraq, and the Uzbeks in Transoxiana. In their works, Safavid literati associated the Safavid dynasty with Tamerlane (d. 1405) in an effort to consolidate their authority in Central Asia.[158] Istanbul responded to these claims and attempted to influence the changing conditions in Central Asia. In 1568–69, Sokullu Mehmed Pasha, the grand vizier to Selim II (r. 1566–74), promoted a project to increase the Ottoman presence in the region. The grand vizier wanted to open a canal between the Don and Volga rivers. The project's aim was to create easier connections between the Ottoman lands and Central Asia. While the plan failed to gain the sultan's approval, this project was one of the earliest attempts to consolidate Ottoman control over trade routes in the area and to establish better connections with the independent Muslim rulers in the region.[159]

Forming political and economic alliances with Central Asian Muslim rulers against the Safavids was the only option for the Ottomans. During the peak of the conflict with the Safavids in the 1580s, the Ottomans continued to assume a pious Sunni Muslim image vis-à-vis their Shi'i Muslim neighbors, the Safavids. Contemporary accounts picked up on this thread and increasingly portrayed the Ottoman sultans as pious rulers. In 1582, at the 50-day imperial circumcision festival, Murad III made a public display of his piety with the circumcision of his sons, and of thousands of orphans and *devşirme* recruits, as well as through the conversion of the members of the Safavid embassy to

---

[156] Baki Tezcan argues that in fact Murad III sought ways to strengthen the absolutist role of the sultan in the Ottoman Empire. For instance, he took over the appointments to the imperial bureaucratic and administrative structures. Murad even introduced new actors to the political stage and tried to curtail the power of the grand viziers. Baki Tezcan, *The Second Ottoman Empire: Political and Social Transformation in the Early Modern World* (Cambridge: Cambridge University Press, 2010), 56, 80, and 95–104.

[157] For a discussion on the Ottoman-Safavid imperial competition and its effects on religious-political sensibilities and the genre of conversion narratives, see Tijana Krstić, "Illuminated by the Light of Islam and the Glory of the Ottoman Sultanate: Self-Narratives of Conversion to Islam in the Age of Confessionalization," *Comparative Studies in Society and History* 51, no. 1 (2009): 35–63.

[158] On Shah Abbas I and his policies and their articulation by Safavid historians, see Newman, *Safavid Iran*, 50–73.

[159] Casale, *The Ottoman Age of Exploration*, 135–6.

Sunni Islam.[160] He commissioned chroniclers and miniaturists to depict and describe these ceremonies, and to articulate his piety.[161] Concomitantly, Ottoman literati highlighted the prestige and political legitimacy of the dynasties of Genghis Khan in Central Asia. The renowned historian Mustafa Ali (d. 1600), for example, compared the universal empires of Tamerlane and Genghis Khan to that of the Ottomans in his monumental world history, the *Künhü'l-Ahbar* (*Essence of History*), written between 1591 and 1599.[162] Mustafa Ali and several other contemporary historians prepared works for the Ottoman sultan in which they attached great importance to the Uzbeks and Crimean Khans because of their descent from Genghis Khan.[163] While the Ottomans could never claim to be descendants of Genghis Khan, they sought to make extensive use of their only plausible connection to Central Asia in this period.

In 1593–94, the historian Talikizade delineated 20 qualities of the Ottoman dynasty that assured its supremacy in the Islamic world. One of these qualities, Talikizade asserts, was the Ottoman dynasty's descent from the Central Asian Turkic ruler Oghuz Khan.[164] These references to the Ottoman lineage going back to Genghis Khan indicate the Ottoman awareness of shifting economic and political networks at the end of the sixteenth century. Ottoman sultans were no longer the Roman emperors; instead, they were pious Sunni Muslims who tried to bolster their connections to Central Asia to maintain the geopolitical status quo and their claim to the leadership of the Muslim world.

Boundaries of the Ottoman World and Ottoman Geographical Knowledge

On the eve of the Lepanto defeat and the conquest of Cyprus, two more *portolan* atlases were produced in Europe for Ottoman court elites. The atlas of Ali Macar Reis, which is today preserved at the Topkapı Palace Library, dates from 1567.[165] Around the same time (ca. 1570), another atlas, the *Atlas-ı Hümayun* (*Imperial Atlas*), was put together for an apparently wealthy and prestigious Ottoman patron. The little that we know about the commissioner and the cartographer confirms that members of the Ottoman court continued to be regarded as prestigious patrons for European maps in this period: a trend that began during the reign of Mehmed II, who commissioned European geographers to prepare paintings

---

160 Derin Terzioğlu, "The Imperial Circumcision Festival of 1582: An Interpretation," *Muqarnas* 12 (1995): 85–6.

161 Ibid.; Christine Woodhead, "Murad III and the Historians of Ottoman Imperial Authority in Late 16th-Century Historiography," in *Legitimizing the Order*, ed. Karateke, 85–98.

162 Fleischer, *Bureaucrat and Intellectual*, 277.

163 Ibid.

164 Necipoğlu, *The Age of Sinan*, 30.

165 Ali Macar Reis, *Atlas*, TKS, Hazine 644.

and translations of important geographical works for him.[166] Reminiscent of the style of the *Walters Sea Atlas*, these two atlases also contain mostly charts of the Mediterranean.[167]

The structure of the three Ottoman atlases from the late sixteenth century—one from the end of the reign of Süleyman I and two from the reign of Selim II—continue to reflect the Ottomans' imperial aspirations and their attempts to be a world power even as imperial conquests were slowing down. Each atlas was prepared for wealthy patrons and each depicts the Mediterranean with greater frequency than those circulating in Europe. Except for the *Walters Sea Atlas*, which contains a chart of the Indian Ocean, the extant atlases do not provide any record of European expeditions across the Atlantic and into the Indian Ocean. This emphasis on the Mediterranean clearly seeks to reaffirm and reiterate to the highest echelons of the Ottoman bureaucracy and administration the centrality of the Mediterranean. Although the sixteenth century was a time of world exploration, the prevailing model of the world empire was still the Roman one. As the conquerors of Byzantium—the New Rome—the Ottomans could not assert universal imperial power without controlling the Mediterranean.

Although the Mediterranean was central to the Ottomans, during the Age of Exploration no imperial power could entirely ignore the New World. Toward the end of the sixteenth century, an anonymous Ottoman geographer compiled the first Ottoman and Muslim account of the New World, *Tarih-i Hind-i Garbi* (*History of the India of the West*). The author was perhaps an Ottoman armchair geographer or a practicing astronomer who worked with an assistant to translate the latest works on the New World. The work relates the history of the first six decades of European discoveries, and also includes information on the natural history, physical geography, and anthropology of the New World. With the help of his assistant, the author of *History of the India of the West* translated brief passages from contemporary Spanish accounts and included them in his work. These passages suggest that the author and his assistant were working with the Italian translations or editions of these books. But the author drew on other sources as well. Among them were medieval Muslim accounts on universal cosmography, such as Zakariya Qazvini's (d. 1283) *'Aja'ib al-Mahlukat* (*The Wonders of Creation*) and Ibn al-Wardi's (d. 1349) *Kharidat al-'Aja'ib wa Faridat al-Ghara'ib* (*Pearls of Marvels and Unique Curiosities*).[168]

166 For differing opinions on the place of production of the *Ali Macar Reis Atlas* and the *Imperial Atlas*, see Goodrich, "Atlas-ı Hümayun: A Sixteenth-Century Ottoman Maritime Atlas Discovered in 1984," *Archivum Ottomanicum* 10 (1985): 83–101; Svat Soucek, "The 'Ali Macar Reis Atlas' and the Deniz Kitabı: Their Place in the Genre of Portolan Charts and Atlases," *Imago Mundi* 25 (1971): 17–27.

167 For further historical analysis of these atlases, see Chapter 3.

168 Thomas Goodrich, *The Ottoman Turks and the New World: A Study of Tarih-i Hind-i Garbi and Sixteenth-Century Ottoman Americana* (Wiesbaden: Harrassowitz, 1990), 20–21, 32–8.

History of the India of the West, therefore, was a very synthetic work. The earliest surviving manuscript is from the late sixteenth century and was presented to Murad III. There are more than 20 manuscript copies of the text from the seventeenth century, including a Persian translation. The author of *History of the India of the West* was obviously keen to create a unique Ottoman account of the New World by combining European and traditional Muslim accounts of universal cosmography and geography. The number of manuscript copies produced evinces the value and popularity of the work among Ottoman geographers and readers in the sixteenth and seventeenth centuries. *History of the India of the West* was among the sources used by the seventeenth-century geographer Katib Çelebi (d. 1657), author of *Cihannüma* (*Views of the World*), a voluminous world geography. *History of the India of the West* was also the fourth book published by the first Ottoman printing press opened by İbrahim Müteferrika (d. 1745) in 1727. Standing at the junction of European and Muslim geographical literature, *History of the India of the West* demonstrates the Ottoman interest in current geographical knowledge on the New World at a time when territorial expansion had halted, and when the empire was going through social, economic, and political crises.[169]

Interest in the lands on the periphery of the Ottoman Empire is evident in another work. At about the same time that *History of the India of the West* was compiled for the Ottoman sultan, Seyfi Çelebi (d. ca. 1590) compiled a work depicting the history, geography, economic life, customs, and rulers of China, India, Transoxiana, and Persia. His account, entitled *Book on the Histories of the Rulers of India and China*,[170] was quite detailed and thorough, yet it was markedly less ornamented in comparison to similar works of the period. To date, historians have been unable to shed light on the author's identity.[171] In the opening lines of his work, Seyfi Çelebi describes himself as a provincial treasurer who was compiling this work on behalf of the reigning sultan, Murad III (r. 1574–95).[172] Seyfi Çelebi's book appeared in Constantinople around the same time that Khitayi's account on China was translated into Ottoman Turkish and *History of the India of the West* was presented to the Ottoman sultan. Seyfi Çelebi opens his account by explaining that "history is such a noble learning [...] that if the rulers learn about the previous rulers and how they ruled, they would become better rulers. Also, if they learn that all these previous rulers had passed away, they would not be so attached to worldly

[169] For the historical significance of *History of the India of the West* and its contribution to the Ottoman understanding of the New World, see Chapter 4.

[170] There are only two extant manuscripts of this work, one in Leiden and the other in Paris. Seyfi Çelebi, *Kitab-i Tevarih-i Padishahan-i Vilayet-i Hindu ve Hitay*, Leiden University Library Cod. 917 (1) Warn.; Paris National Library Suppl. Turc. No. 1136. Joseph Matuz prepared the facsimile edition of the Leiden copy in 1968: Joseph Matuz, ed., *L'Ouvrage de Seyfi Çelebi: Historien Ottoman du XVIe Siecle* (Paris: Dépositaire, 1968).

[171] Franz Babinger, *Die Geschichtsschreiber der Osmanen und Ihre Werke* (Leipzig: Otto Harrassowitz, 1927), 69; Matuz, ed., *L'Ouvrage de Seyfi Çelebi*; Christine Woodhead, "Seyfi," *EI*², last accessed via Brill Online. University of Pittsburgh. 7 November 2011.

[172] Matuz, ed., *L'Ouvrage de Seyfi Çelebi*, 163.

affairs and would be more concerned with the adulation of God."[173] In the beginning of the book, he addresses the history, geography, society, and rulers of China in only 20 pages. In the rest of the work, he focuses on the history and genealogies of the rulers of India, Transoxiana, and Persia. He pays particular attention to the Sunni Muslim rulers in Central Asia and emphasizes their genealogies going back to Genghis Khan. Whoever the author was, he clearly aspired to impress the Ottoman court with his command of Turkish, his eloquent language, and his knowledge of the history of Muslim rule in Central Asia. Published in an era when Ottoman influence in Central Asia was a growing concern, one can surmise that the author hoped to gain the sultan's patronage.

Also built in this period was the first Ottoman observatory. Murad's grand vizier Sokullu Mehmed Pasha convinced the sultan to allow Takiyüddin (d. 1585), the *müneccimbaşı* (court astrologer),[174] to establish an observatory in Galata in 1579.[175] Although the observatory was very short lived, its instruments were state-of-the-art and very similar to the instruments used by Takiyüddin's contemporary, the Danish astronomer Tycho Brahe.[176] However, the observatory was razed in 1580 after Şeyhülislam Ahmed Şemseddin Efendi, son of Ebussu'ud Efendi, convinced Murad III that the observation of the stars brought bad luck to the Ottomans in their struggle against the Safavids.[177] Not only the observations and the up-to-date instruments themselves but also the fact that the Ottoman state encouraged the construction of a space for both creating knowledge through observation via the use of the latest instruments and assembling existing geographical knowledge, was an important stage for the development of an Ottoman geographical consciousness.[178] The observatory offered a space for those who were interested in astronomy and

---

173 Ibid.

174 For the *müneccimbaşı*'s role in the Islamic societies, see George Saliba, "The Role of the Astrologer in Medieval Islamic Society," *Bulletin D'Études Orientales* 44 (1992), 45–67.

175 For further information on this observatory and its foundation, see Adnan Sayılı, *The Observatory in Islam and Its Place in the General History of the Observatory* (Ankara: TTK, 1960), 290–92. See also D. A. King, "Taki al-Din," *EI²*, last accessed via Brill Online. University of Pittsburgh. 7 November 2011; J. H. Mordtmann, "Das Observatorium des Taqi ed-Din zu Pera," in *Astronomische Instrumente Publikationen 1918–1925*, vol. 4 of *Arabische Instrumente in orientalischen Studien*, ed. Fuat Sezgin (Frankfurt: IGAIW, 1991), 281–95; Avner Ben-Zaken, *Cross-Cultural Scientific Exchanges in the Eastern Mediterranean, 1500–1600* (Baltimore: Johns Hopkins University Press, 2010).

176 Seyyid Lokman's *Alatü'r Rasadiyye li Zijüş'-Şahinşahiyye* (Astronomical Instruments for the Imperial Astronomical Tables) lists the instruments that were used in his observatory. On this work and its authorship, see Baki Tezcan, "The Multiple Faces of the One: The Invocation Section of Ottoman Literary Introductions as a Locus for the Central Argument of the Text," *Middle Eastern Literatures* 12 (2009): 27–41.

177 Sayılı, *Observatory*, 290–92; D. A. King, "Taki al-Din"; Mordtmann, "Das Observatorium des Taqi ed-Din zu Pera," 83.

178 For further examples of this phenomenon in early modern Europe, see David Turnbull, "Cartography and Science in Early Modern Europe: Mapping the Construction of Knowledge Spaces," *Imago Mundi* 48 (1996), 5–24.

geography to get together and engage in scientific activities. The geographical accounts and the observatory informed each other. For instance, one of the central features of the observatory, which was depicted in Seyyid Lokman's *Şahanşah-name*, was a terrestrial globe. In the miniature depiction, the globe, which stands in the middle of the picture, centers on the Ottoman domains in the Mediterranean and Southeast Europe. Although the dimensions of Africa are exaggerated, the globe also includes parts of Asia, Europe, and the coastline of South America to the Atlantic Ocean (Plate 1).[179]

Constructed during the peak of the Ottoman-Safavid conflict and the stalemate in the Mediterranean, this observatory along with the geographical works indicate the Ottomans' awareness of the expanding world and of shifting economic and political networks at the end of the sixteenth century. The Ottoman sultans' claims to be the Roman emperors and pious Sunni Muslim rulers of the universe continued to be articulated in the historical and geographical literature of the time. In addition to the above-mentioned geographical works, copious amounts of world histories, histories of the universe, and cosmographies prepared by Ottoman intellectuals for Ottoman sultans appeared at this time.[180] The steady increase in the number of translations and compilations of traditional cosmographical accounts or book of wonders in Arabic and Persian, as well as pilgrimage and travel accounts from this period, also deserves attention here.[181] As the Ottoman court welcomed these traditional geographical accounts, their patrons encouraged the Ottoman geographers to find new ways to articulate the concerns of the period. The Ottoman geographers prepared accounts on the different regions of the world and on the Ottoman territories. Their works had a significant influence on Ottoman politics and aided in the Ottoman court's quest for imperial legitimacy. That the maps, *portolan* charts, navigational guides, and travel accounts validating the court's assumptions and ambitions were produced by intellectuals who worked independently of an institutional framework gave the works considerable symbolic and political importance. Depending on which member of the court commissioned these accounts, their content served the political or economic aims of one party or the other at the Ottoman capital. Ottoman geographers were thus suitably rewarded for their contributions by the sultan and his court elites. By viewing these works in their proper historical context, Ottoman geographers can now be seen as important players in the development of the Ottoman imperial consciousness, and as vital contributors to the intellectual life that shaped and was shaped by the currents of the empire and globe.

[179] Adnan Sayılı, *The Observatory in Islam*, 294–5.

[180] On the increasing number of Ottoman illustrated histories from the second half of the sixteenth century and their role in shaping the political agendas of the period, see Emine Fetvacı, *Picturing History at the Ottoman Court* (Bloomington and Indianapolis: Indiana University Press, 2013).

[181] As opposed to less than five translations of cosmographical accounts from the fourteenth to the sixteenth century, more than 20 compilations and translations on this topic can be listed from the sixteenth century. İhsanoğlu, *Osmanlı Coğrafya Literatürü Tarihi*, vol. 1, 17–83.

Chapter 2
Mapping and Describing Ottoman Constantinople

> Let no one doubt that he [Mehmed II] is by right the emperor of the Romans. For he is emperor who by right possesses the seat of the empire, but the seat of the Roman Empire is Constantinople: thus he who by right possesses this city is the emperor. But it is not from men but from God that you, thanks to your sword, have received this throne [...] And he who is and remains emperor of the Romans is also emperor of the entire earth.[1]

Constantinople had held an important place in eschatological discussions and political ideology since its foundation in the fourth century. Under Byzantine rule, the city flourished and became a cosmopolitan center of trade and intellectual and artistic achievement. On May 29, 1453, Ottoman sultan Mehmed II conquered the Byzantine capital. This long-awaited conquest, one of the major aspirations for Muslim rulers since the seventh century, ended not only Byzantine control of the city but also of its empire. A little more than a decade later, in 1466, a Byzantine scholar and philosopher, George Trapezountios, wrote a letter to Pope Paul II. In this letter and other writings, Trapezountios maintained that Mehmed II, who had been called by God to rule the entire world, would soon conquer Rome. But he also predicted that Mehmed's domains would disintegrate soon thereafter if he did not convert to Christianity: "God will have one universal kingdom, one church, and one faith. If you accept this opportunity and confess the true religion, He will make your progeny eternal and universal autocrats; if not, He will destroy your domination once you have established political unity."[2]

For Mehmed II, the conquest of Constantinople was of immense strategic and symbolic importance. Mehmed had struggled to exert his authority ever since he first ascended to the throne as a 12-year-old boy in 1444 when his father, Murad II, abdicated in his favor. After a fire devastated the capital of Edirne and a Janissary rebellion broke out, the grand vizier Çandarlı Halil decided to invite Murad II back onto the throne. When the now 19-year-old Mehmed reclaimed the throne in 1451, the ruling elites of his father immediately challenged his

[1] George Trapezountios's letter to the Pope, cf. Franz Babinger, *Mehmed the Conqueror and His Time*, trans. Ralph Manheim, ed. William Hickman (Princeton: Princeton University Press, 1978), 251.

[2] John Monfasani, *George of Trebizond: A Biography and a Study of His Rhetoric and Logic* (Leiden: Brill, 1976), 132.

authority.[3] Only an impressive conquest would legitimize his rule and secure his position. Constantinople was the perfect place for such a conquest. It was an urban island—a red apple—in the middle of his domains, and its conquest would result in both material and symbolic gains. Constantinople was not only a center of international trade,[4] but also the capital of the Eastern Roman Empire and the center of Eastern Christianity. Its eventual conquest provided Mehmed and his ruling elite with an imperial capital, and allowed them to appropriate the power of the Byzantine legacy. While the cities of Bursa and Edirne had previously served as Ottoman capitals, Constantinople, as the New Rome, had been the center of the Roman Empire and consequently the world. Conquering such a city gave the Ottoman ruling elites a chance to emulate a great empire while establishing their own—they took full advantage of this opportunity.[5]

This chapter explores the ways in which Ottoman geographers articulated the centrality of Constantinople in the service of the imperial court's political agenda. It discusses the role of Constantinople in the construction of Ottoman political ideology and geographical works, from its beginnings as the New Rome in Byzantine times to the point when it became an important component of the Ottoman court's imperial vision and geographical consciousness. A close examination of historical accounts and geographical works prepared for the Ottoman sultans in Europe and the Ottoman Empire starting in the fifteenth century—such as Francesco Berlinghieri's treatise on Ptolemy's *Geographia*, Matrakçı Nasuh's *Beyan-ı Menazil*, and Mustafa b. Ali's *Public Instructiuons*—reveals that the Ottoman court employed geographical works in the service of its imperial claims. They also show that Mehmed II was the first Ottoman sultan who used geographical works to articulate his imperial vision. This was a legacy he left to future Ottoman sultans.

---

3 Imber, *The Ottoman Empire*, 26–7. On Mehmed II's reign, see Babinger, *Mehmed the Conqueror*; Halil İnalcık, "Mehmed the Conqueror (1432–1481) and His Time," *Speculum* 25 (1960): 408–27, reprinted in Halil İnalcık, *Essays in Ottoman History* (Istanbul: Eren, 1998): 87–110; Halil İnalcık, "The Policy of Mehmed II towards the Greek Population of Istanbul," *Dumbarton Oaks Papers* 23–4 (1969–70): 231–49; Michael Doukas, *The Decline and Fall of Byzantium to the Ottoman Turks*, trans. H. J. Magoulias (Detroit: Wayne State University Press, 1975); Colin Heywood, "Mehmed II and the Historians: The Reception of Babinger's *Mehmed der Eroberer* during Half a Century," *Turcica* 40 (2008): 295–344.

4 Despite the fact that Constantinople hardly contributed to economic production, it was nonetheless an important *entrepôt* in international trade as a consumer city. On Constantinople's role in international trade, see Edhem Eldem, "Istanbul: From Imperial to Peripheralized Capital," in *The Ottoman City between East and West: Aleppo, Izmir, and Istanbul*, ed. Eldem, Goffman, and Masters (Cambridge: Cambridge University Press, 1999), 135–206.

5 In conquering and making Constantinople their capital city, the Ottomans faced the significant challenge of figuring out how to use this great Byzantine legacy to their advantage. For a discussion on the selective appropriation of Byzantine Constantinople's imperial legacy in the construction of Ottoman Constantinople, see especially Kafescioğlu, *Constantinopolis/Istanbul*, 53–143.

The capture of Constantinople irrevocably changed the Ottomans' perception of their own state. Constantinople provided them with a new political center, where the Ottomans formulated a new, centralized state vision. The semi-nomadic, ghazi state that was stationed in Bursa and Edirne gave way to a new model of empire, one that was sedentary and centralized, and which would be further refined in the sixteenth century. Geographical works assisted in this process until the fifteenth-century political vision that placed Constantinople at the center of the Ottoman Empire evolved into the sixteenth-century perception that Constantinople stood at the center of the entire universe, radiating Ottoman power throughout the world and beyond.

Where Is the New Rome?

In 324, Roman Emperor Constantine the Great established his new capital on the site of the ancient city of Byzantium. According to contemporary accounts, he initially intended to found a city on the plains of Ilion at the foot of Troy. But, according to the Byzantine lore that these accounts relate, he changed his mind after God appeared to him in a dream and told him to build his city in Byzantium, an ancient town on the Bosphorus. Believing that he had been consecrated by God, Constantine set to work on turning this ancient town into a new capital.

Constantine wanted his new city to resemble Rome as much as possible. He established a senate by decree, invited influential Romans to settle in the city, erected imperial monuments, and imported columns and statues from the provinces to symbolize his divine and imperial authority.[6] Soon after its foundation, Constantinople became a vibrant political and economic center. The city's fortifications proved to be formidable barriers against the Vandals and Goths who would sack Rome in the fifth century. Over the next few centuries, Constantinople would survive attacks by the Sassanids of Persia, Muslim armies, the Normans, and the Turks.

For Byzantine rulers, Constantinople was the New Rome, the unquestionable center of the united Roman Empire, now ruled by Byzantine emperors. The city held an important place in the political, economic, and religious life of the Mediterranean, and quickly became a magnet for artists, craftsmen, traders, and intellectuals, remaining a vibrant capital until the thirteenth century. As the political center of the Byzantine Empire, Constantinople was a constant target of attack by armies from both the East and West. It managed to survive all military attacks, however, until 1204. Not only was Constantinople a political center, but it was also an economic hub. The city was a major *entrepôt* in the eastern Mediterranean, connecting the Silk Road to trading centers in Europe and the Mediterranean. After being adorned with the Hagia Sophia, the city became an

---

6 William Stearns Davis, ed., *Readings in Ancient History: Illustrative Extracts from the Sources, Vol. 2, Rome and the West* (Boston: Allyn and Bacon, 1912–13), 295–6; Stéphane Yerasimos, *Constantinople: Istanbul's Historical Heritage* (Paris: Konemann, 2005), 28.

important center for Eastern Christianity as well. By assuming these varied roles, Constantinople challenged the prominence of ancient political, economic, and religious centers, most notably Alexandria, Jerusalem, and Rome. In Byzantine political thought, Rome still held a prestigious position as the capital of the Roman Empire. Constantinople, however, loomed even larger as a new city and as the religious center of the new imperial faith.[7]

In 800, Pope Leo III crowned Charlemagne emperor of the Romans and put the Byzantines' sway to the test. Historical literature has repeatedly discussed this event and its ramifications for European political history. These discussions focus on the coronation's impact on Frankish rule in Europe, and on Charlmagne's role in shaping the imperial model for future rulers in Europe.[8]

But how did this event affect Constantinople and its place in European political thought? The revival of the Roman Empire in the West was undoubtedly a direct challenge to the tenet of Byzantine ideology that the eastern and western parts of the Roman Empire would one day be united.[9] But while the Pope aspired to reunite the Roman Empire under Christian rule, Charlemagne's political authority remained limited to certain territories in Germany—his sovereignty over other parts of Europe was disputable.[10] It was so disputable, in fact, that Charlemagne rejected the Pope's portrait of him as the embodiment of a universal rulership that was centered in Rome, and instead adopted the Byzantine imperial model, choosing to rule his empire from Aachen, which he treated as his new Rome.[11] It was only toward the end of the tenth century that Otto II of the Saxon dynasty embraced the title *Romanorum Imperator Augustus* (August Emperor of the Romans). His son and successor, Otto III, further strengthened the image of Rome as the center of a Christian empire. His seal was encircled by the words "*Renovatio imperii Romanorum*" (Renewal of the empire of the Romans). But it was not until the crusaders invaded Constantinople in 1204 that Rome regained its self-appointed position in Latin Christendom as the undisputed political and religious center of the world.[12]

---

7 D. M. Nicol, "Byzantine Political Thought," in *The Cambridge History of Medieval Thought c. 350–c. 1450*, ed. J. H. Burns (Cambridge: Cambridge University Press, 1988), 52–3; Joseph Canning, *A History of Medieval Political Thought 300–1450* (London: Routledge, 1996), 5.

8 For a general discussion of the coronation and its impact on Frankish rule in Europe, see Janet Nelson, "Kingship and Empire," in *The Cambridge History of Medieval Thought c. 350–c. 1450*, ed. J. H. Burns (Cambridge: Cambridge University Press, 1988), 230–34. For Charlemagne and his role as a model for imperial rulership in sixteenth century Europe, see Yates, *Astraea*, 4–5; Kleinschmidt, *Charles V*, 82.

9 Nicol, "Byzantine Political Thought," 58.

10 Yates, *Astraea*, 2–3.

11 Canning, *A History of Medieval Political Thought 300–1450*, 69.

12 Ibid., 74–8. Even in the apocalyptic traditions, Rome replaced Constantinople after 1204. In these scenarios, Rome had a double role: it was both the gateway to paradise, and the arena for the Antichrist that had to be chastised, even destroyed, for its wickedness. Marjorie Reeves, "The Medieval Heritage," in *Prophetic Rome in the Renaissance Period*, ed. Marjorie Reeves (Oxford: Clarendon Press, 1992), 3–21.

Constantinople also held a significant place in Muslim political ambitions and religious discussions during this period. As the capital city of the Byzantine Empire and of Eastern Christianity, its conquest by the Muslim armies was an apocalyptic event.[13] The Muslim apocalyptic tradition had associated the destruction of Constantinople with the end of time itself,[14] and a *hadith* had prophesied its fall to Muslim armies.[15] Inspired by these speculations and ambitious to defeat their rival to the north, Caliphate armies had laid siege to the city 12 times over the course of seven centuries before its conquest by the Ottomans in 1453. Although over time the apocalyptic traditions predicting the destruction of the city had eventually lost currency and been forgotten, they resurfaced during the Ottoman conquest of the city.[16]

All Roads Lead to Constantinople: The New Rome in Pre-Ottoman Geographical Traditions

Since the fifth century, Byzantine emperors saw themselves as the sole rulers of the Roman Empire, and considered Constantinople its new capital city. It was the New Rome, the new center of the Roman Empire. Before long, the city became the focus of political and eschatological discussions both in the Muslim and Christian worlds. It rivaled other political and religious centers like Antioch, Jerusalem, Rome, and Mecca. This rivalry generated a geographical and cartographical tradition in the Byzantine Empire,[17] where emperors patronized geographical works to mirror their

---

[13] According to David Cook, Muslim apocalyptic is a genre of literature presented in the form of a *hadith* purporting to convey information about the period of time leading up to the end of the world, or to give the impression that the historical events occurring in the present are an integral part of this finale. David Cook, *Studies in Muslim Apocalyptic* (Princeton: Darwin Press, 2002), 1.

[14] Both Muslims and Christians derived their imagery from the fall of Babylon described in Revelation 18:1–24. This is not surprising since Constantinople was considered to be the New Rome, and the original apocalypse in Revelation was directed against the pagan idol-worshipping city of Rome (code-named Babylon). Cook, *Studies in Muslim*, 54–62.

[15] It is still disputable if the *hadith* is accurate (*sahih*). The so-called *hadith* reads: "Verily you shall conquer Constantinople. What a wonderful leader will her leader be, and what a wonderful army will that army be!"

[16] William J. Hamblin, "Arab Perspectives on the Fourth Crusade," in *The Fourth Crusade: Event, Aftermath, and Perceptions* (Papers from the Sixth Conference of the Society for the Study of Crusades and the Latin East, Istanbul, Turkey, 25–29 August, 2004), ed. Thomas F. Madden (Burlington, VT: Ashgate, 2008), 167–79.

[17] Our knowledge of Byzantine geographical tradition is very fragmented. The earliest examples of Byzantine geographical works—two sixth-century map mosaics in Madaba and Nicopolis, and the geographical works of Cosmas Indicopleustes, a merchant from Alexandria—are heavily embedded in biblical concepts. For a discussion on the Madaba map's cartographic context and its comparison to contemporary mosaics, see P. D. A. Harvey, "The Cartographic Context of the Madaba Map," in *The Madaba Map Centenary 1897–1997: Travelling through the Byzantine Umayyad Period*, ed. Michele Piccirillo and Eugenio Alliata (Jerusalem: Studium Biblicum Franciscanum, 1999), 103–7. The world map that

ambitions for the city. An anonymous treatise, *Urbs Constantinopolitana nova Roma*, which was prepared for and dedicated to the Byzantine Emperor Theodosius II (r. 408–50), is a good starting point to explain how Byzantine emperors endeavored to articulate the importance of the city by employing available geographical traditions. Deeply rooted in similar treatises on Rome, *Urbs Constantinopolitana* lists the 14 districts of the city, describes its monuments, and names its civil servants.[18] Unfortunately, no maps accompany this text;[19] however, it clearly illustrates how Byzantine rulers replicated Rome in their new city, with its seven hills, 14 districts, and a similar number of Christian temples, each of which was built as a counterpart to one of the temples in Rome.[20] Byzantine rulers were not alone in their patronage of Constantinople's geographical representations. According to Charlemagne's courtier and biographer, Einhard, the Carolingian ruler had three silver tables adorned with maps in his possession. One of these tables depicted the whole world, while the other two pictured Rome and Constantinople.[21]

Despite the efforts of Byzantine emperors, many contemporary authors and geographers either completely ignored the existence of Constantinople or adopted a negative attitude toward it. For example, a fourth-century Roman historian, Ammianus Marcellinus, made no reference to Constantinople in his history of Rome. Marcellinus, a native and resident of Antioch, was writing in Rome for Rome—the intellectuals of both cities resented Constantinople's ascendancy in political and religious life.[22]

---

was prepared for Theodosius II (r. 408–50) is another example of the Byzantine interest in geographical works. Although the map itself has been lost, the poem in Latin attached to it explains that the two cartographers were commissioned to update an older world map, possibly of Roman origin. In addition, manuscript collections in different European libraries indicate that several copies of Ptolemy's works were produced in Constantinople in the thirteenth century. O. A. W. Dilke, "Cartography in the Byzantine Empire," in *The History of Cartography, Vol. 1, Cartography in Prehistoric, Ancient, and Medieval Europe and the Mediterranean*, ed. J. B. Harley and David Woodward (Chicago: University of Chicago Press, 1987), 258–76.

[18] *Geographi Latini Minores*, ed. Alexander Riese (Heilbronn: Apud Hennineros Fratres, 1878), 133–9.

[19] Dilke, "Cartography in the Byzantine Empire," 259.

[20] A similar (but much longer and more detailed) text from the fifth century by Marcellinus also lists the monuments and the civil servants of the city. Dilke, "Cartography in the Byzantine Empire," 259.

[21] David Woodward, "Medieval Mappaemundi," in *The History of Cartography, Vol. 1, Cartography in Prehistoric, Ancient, and Medieval Europe and the Mediterranean*, ed. J. B. Harley and David Woodward (Chicago: University of Chicago Press, 1987), 302–3. According to Harvey, there might be a continuing tradition of drawing the plans of these two cities in the medieval period. Extant medieval depictions of these cities, however, give no hint of classical antecedents. P. D. A. Harvey, "Local and Regional Cartography in Medieval Europe," in *The History of Cartography, Vol. 1, Cartography in Prehistoric, Ancient, and Medieval Europe and the Mediterranean*, ed. J. B. Harley and David Woodward (Chicago: University of Chicago Press, 1987), 467.

[22] Gavin Kelly, "The New Rome and the Old: Ammianus Marcellinus' Silences on Constantinople," *Classical Quarterly* 53, no. 2 (2003): 588–607.

Marcellinus was not alone in his treatment of Constantinople. In the world maps prepared in medieval Europe, Constantinople was either given no representation or cast as secondary in importance to Jerusalem.[23] Among the most prominent examples of this cartographic tradition were the Ebstorf map (ca. 1235) and the Hereford map (ca. 1290), both of which were geographical representations of the world, as well as religious iconographies. While the Ebstorf map was displayed in a convent in Germany, the Hereford map was kept in Hereford Cathedral.[24] Both of these maps were modified versions of the T-O map (Figure 2.1).[25]

The map from Hereford (Figure 2.2), which is the only survivor of the two, represents the view of the world from a Christian perspective in the thirteenth century, with depictions of mythical creatures and abnormal people placed alongside new information gathered from pilgrimages, crusades, and commercial travels.[26] The eastern Mediterranean, including the Anatolia and Syria, are at the center of the map. Babylon is at the top. Along the right edge, the cartographer depicts the route followed by Israelites in their exodus from Egypt. North and East Asia are in the upper left corner. The lower left and upper right corners chart the "islands" that are home to the evil offspring of Cain. At the center of the map is Jerusalem; there is a crucifix above it.[27] In addition to these Christian iconographies, the map depicts mythical creatures such as a mermaid in the Mediterranean, an Alerion in Asia, a centaur in Egypt, dragons in India, and huge ants in Africa.[28] It portrays Constantinople as almost identical to Rome but insignificant, and of no more merit than Alexandria.[29]

Portolan charts and atlases of the medieval period complied with the *mappaemundi* with regard to the representations of Constantinople. *Portolan*

---

23 World maps prepared in this period tried to squeeze three of the four known continents into the shape of a circle. Larger maps encompassed topography, ethnography, and natural science, and in some cases Christian historiography and even literary—or at least legendary—history. Therefore, a large majority of medieval maps are to be seen as works of art rather than sources of information. Rudolph Simek, *Heaven and Earth in the Middle Ages: The Physical World Before Columbus*, trans. Angela Hall (Woodbridge, UK: Boydell & Brewer, 1996), 41–2.

24 P. D. A. Harvey, *Medieval Maps* (Toronto: University of Toronto Press, 1991), 29–30.

25 In this map projection, Asia occupies the upper half and is separated from Africa by the Nile and from Europe by the Don; the Mediterranean separates Europe from Africa. Therefore, the lands acquire the shape of a T, while a circumfluent ocean—the O—surrounds these lands. Norman Thrower, *Maps and Civilization: Cartography in Culture and Society*, 2nd ed. (Chicago: University of Chicago Press, 1999), 42.

26 Thrower, *Maps and Civilization*, 44.

27 For a detailed description of the map, see Scott D. Westrem, "Making a Mappamundi: The Hereford Map," http://cartographic-images.net/Cartographic_Images/226_The_Hereford_Mappamundi.html.

28 Richard de Bello, "The Hereford Mappamundi," http://cartographic-images.net/Cartographic_Images/226_The_Hereford_Mappamundi.html.

29 P. D. A. Harvey, *Medieval Maps*, 7–39.

Fig. 2.1 Earliest printed example of a T-O map by Günther Zainer, *Etymologiae of Isidore of Seville.* Isidori Iunioris Hispalensis Episcopis Epistola, Augsburg, 1472. The Harry Ransom Center, University of Texas at Austin, Kraus 13. Courtesy of the University of Texas at Austin.

charts from the late thirteenth century to the 1500s largely covered the Mediterranean and the Black Sea. They sometimes contained charts of the Atlantic coast of Europe and Africa as well.[30] Italian workshops and Catalans dominated the market for *portolan* charts.[31] The charts provided detailed and mostly accurate renditions of the Mediterranean coastline and the islands. Constantinople as a major trade and religious center always found representation in the charts of the

---

[30] For a reliable survey of the medieval *portolan* charts and atlases, see Tony Campbell, "Portolan Charts from the Late Thirteenth Century to 1500," in *The History of Cartography, Vol. 1, Cartography in Prehistoric, Ancient, and Medieval Europe and the Mediterranean*, ed. J. B. Harley and David Woodward (Chicago: University of Chicago Press, 1987): 371–464.

[31] For a discussion on the origins of the medieval *portolan* charts and atlases, see ibid., 388–90.

Fig. 2.2 Hereford World Map, 1290. Hereford Cathedral Library. At the center of the map is Jerusalem. Courtesy of Hereford Cathedral Library.

eastern Mediterranean and the Black Sea. While the cities of Genoa and Venice, which produced most of the Italian *portolans*, were occasionally illustrated in detail, Constantinople was almost always simply indicated with a flag signaling Byzantine rule.

Similarly, world maps drawn by medieval Muslim scholars also downplayed the importance of Constantinople. Drawing mainly upon the works of Aristotle and Ptolemy and their own intensive observations of the heavens, scholars of the

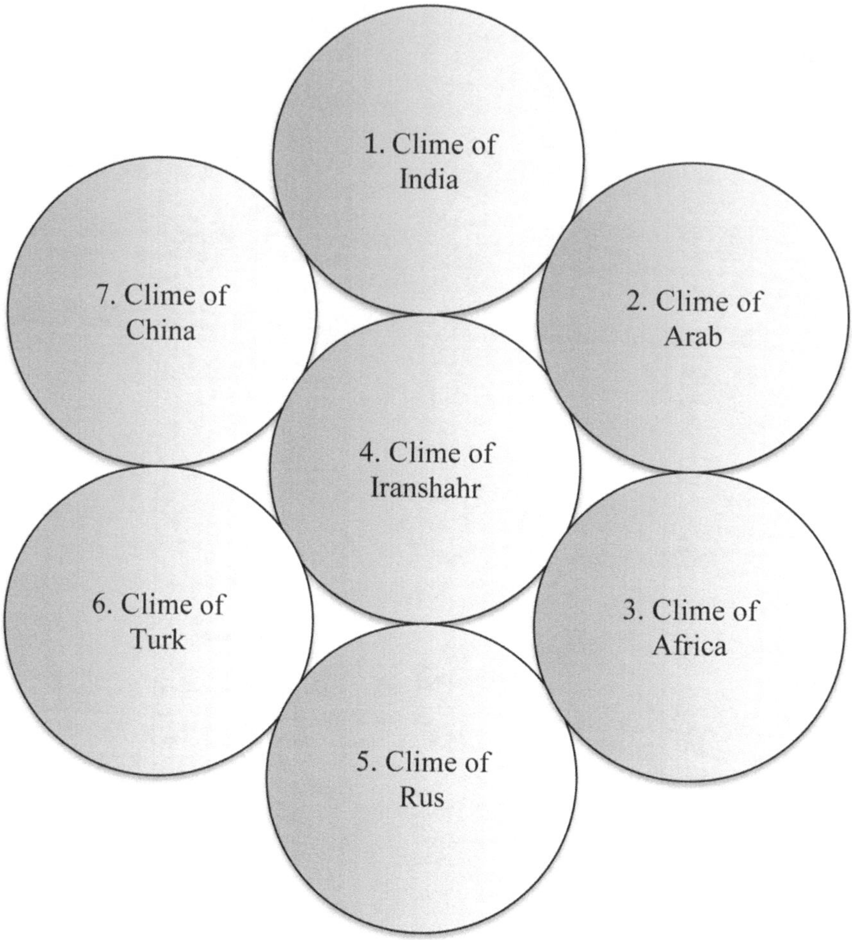

Fig. 2.3 The Seven Regions according to early Muslim scientists. Modeled after *Mujmal at-Tawarikh wa'l-Qisas: Eine persische Weltgeschichte aus dem 12. Jahrhundert*, ed. Seifeddin Najmabadi and Siegfried Weber (Edingen-Neckarhausen: Deux Mondes Publishing, 2000), 367/fol. 263a.

Muslim world created their own cartographical traditions. The earliest world maps drawn by Muslim geographers divided the inhabited world into seven climes. Geographers were influenced by the Persian tradition that divided the world into seven circular regions (*kishvar*) (Figure 2.3) and the Ptolemaic theory that divided the world into seven climes.

The model became particularly dominant after the foundation of the *Bayt al-Hikma* (School of Wisdom) in Baghdad in the ninth century, when scholars started translating Ptolemy's works.[32] In this model, six regions encircle the fourth and central region. While Persia was originally taken as the center in Persian models, later versions of the model usually placed Damascus or Baghdad at the center of the world. In this scheme, the Byzantine domains and Constantinople were part of the fifth region.[33] Later on, as a result of the influence of astronomers, medieval Muslim scientists adopted a new model where the seven regions were situated in parallel below the equator line (Figure 2.4).

Manuscript copies of Al-Idrisi's (d. 1165) geographical account *Nuzhat al-Mushtaq* (*The Book of Pleasant Journeys*) from 1154 contain a number of examples of this model. This geographer from Ceuta was the founder of the Ptolemaic school of geography in the Muslim world. Commissioned by the Norman King Roger II of Sicily, *Nuzhat al-Mushtaq* today survives in 13 manuscript copies.[34] Eight of these copies include regional maps, and six copies contain a small world map.[35] While the cartographer of these maps remains unknown, the maps clearly illustrate the world view of Muslim geographers in the pre-Ottoman period. The extant circular world map from *Nuzhat al-Mushtaq* divides the world into seven climates. The map is south oriented, taking the Ka'ba as the center of the world. The cartographer clearly marks Sicily, Florence, and Venice, but Constantinople does not appear on the map.[36]

---

32 Ptolemy's *Geographia* and *Almagest* had a profound effect on Muslim geographical knowledge. After these works were translated from Greek into Syriac and Arabic in the ninth century, Muslim geographers repeatedly composed treatises on Ptolemy's works, tested his findings on the coordinates of the world, and drew maps by following his instructions. Ahmet Karamustafa, "Cosmographical Diagrams," in *The History of Cartography, Vol. 2, Bk. 1, Cartography in the Traditional Islamic and South Asian Societies*, ed. J. B. Harley and David Woodward (Chicago: University of Chicago Press, 1992), 76–7. See also, Andre Miquel, "Iqlim," *EI²*.

33 One of the earliest examples of this model was drawn by al-Biruni. Ahmet Karamustafa, "Cosmographical Diagrams," 77–80; David A. King and Richard P. Lorch, "Qibla Charts, Qibla Maps, and Related Instruments," in *The History of Cartography, Vol. 2, Bk. 1, Cartography in the Traditional Islamic and South Asian Societies*, ed. J. B. Harley and David Woodward (Chicago: University of Chicago Press, 1992), 189–203.

34 Al-Idrisi's *Nuzhat al-Mushtaq* also found its way into some of the libraries in Istanbul. Today, there are three extant copies of the work in Istanbul. Maqbul Ahmad, "Cartography of al-Sharif al-Idrisi," in *The History of Cartography, Vol. 2, Bk. 1, Cartography in the Traditional Islamic and South Asian Societies*, ed. J. B. Harley and David Woodward (Chicago: University of Chicago Press, 1992), 173–4.

35 Ibid., 156–74; Sonja Brentjes, "al-Idrisi," in *Medieval Science, Technology, and Medicine: An Encyclopedia*, ed. Thomas Glick, Steven J. Livesey, and Faith Wallis (New York: Routledge, 2005), 261. For a discussion of the cartographer of these maps and whether it was al-Idrisi or Roger II or another cartographer, see Jeremy John and Emilie Savage-Smith, "The Book of Curiosities: A Newly Discovered Series of Islamic Maps," *Imago Mundi* 55 (2003): 7–24.

36 For a discussion of al-Idrisi's maps and their meaning for the Muslim understanding of boundaries and frontiers in the medieval period, see Ralph Brauer, "Boundaries and Frontiers in Medieval Muslim Geography," *Transactions of the American Philosophical Society* 85, no. 6 (1995): 3–73.

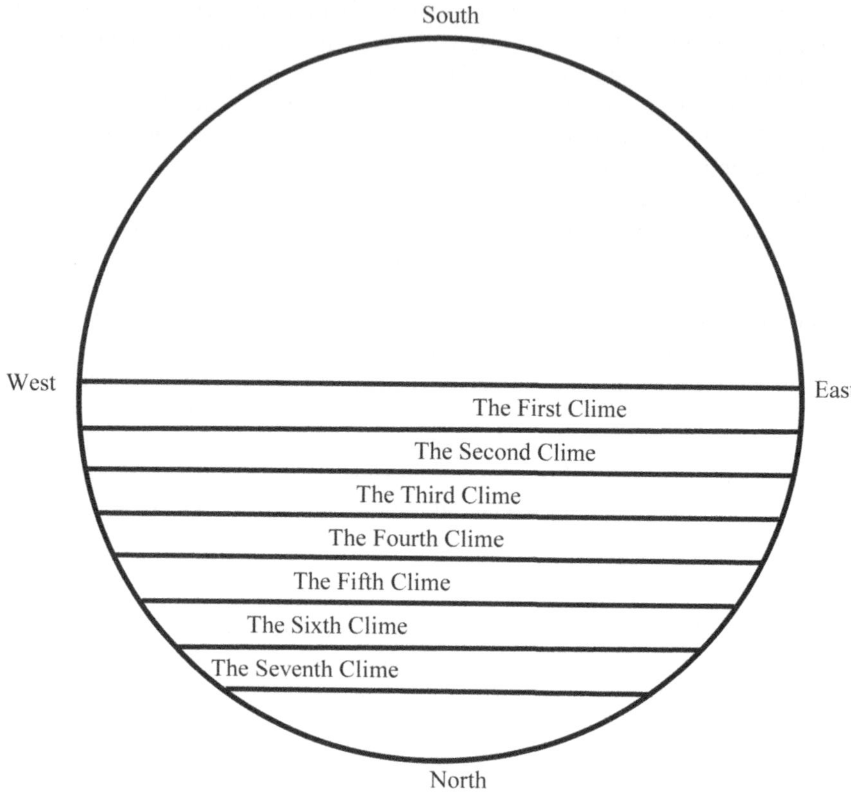

Fig. 2.4 The world according to early Muslim scientists. Modeled after al-Idrisi's circular world map, ms. Oxford, Bodleian, Pococke 375, 3b–4a.

When analyzed together, the geographical works produced in Europe and the Muslim world were not so very different from each other during the pre-Ottoman period. Not only did Islam's and Christianity's sacred texts draw upon the same Judaic sources and share the assumption of a single Creator, but Muslim scholars were also influential in transmitting ancient Hellenistic geographical knowledge to medieval Europe through cultural centers in Spain.[37] The maps that were prepared to satisfy religious claims portrayed the earth of antiquity, where the religious centers of Jerusalem and Mecca still held more importance than political centers such as Constantinople and Rome.[38] The silence about Constantinople on both Muslim and European maps also speaks to resentment in both Europe and the Muslim world against the religious and political importance of the city.

[37] Cosgrove, *Apollo's Eye*, 70.

[38] Natalia Lozovsky, *"The Earth is Our Book": Geographical Knowledge in the Latin West ca. 400–1000* (Ann Arbor: University of Michigan Press, 2000), 139–52.

It is impossible to know the extent to which the Ottomans were aware of these cartographical traditions by the time they conquered Constantinople in 1453. While it is well known that Mehmed II was fascinated with geographical works—especially with maps—it is difficult to determine just how aware he was of the latest developments in geography and of the widespread resentment against Constantinople.[39] Mehmed's conquest of Constantinople brought the Ottomans into contact with a new group of people who had a variety of political, religious, and intellectual traditions, which the Ottomans eventually adopted in the formulation of their own imperial ideology. Soon after the conquest, Mehmed II launched an architectural, social, and political campaign to change the outlook of Constantinople. Geographical works played a vital role in this enterprise.

Mehmed the Conqueror: Constantinople as the Center of the Empire

By the time Mehmed II and his army completed the conquest of Constantinople, the city was in ruins. It had already lost its former cultural and political splendor after the Fourth Crusade in 1204. Between 1204 and 1261, when Constantinople was under Frankish rule, rapacious crusaders stripped the city of its riches and seriously challenged the territories of the Byzantine Empire. In the process, they shattered Byzantine political ideology, which had connected the destiny of the city and the empire. In the apocalyptic discourses circulating in the Byzantine Empire, the city's and the empire's fates were entwined—both were destined to last until the end of the world and the end of time. Constantinople's destruction would occur during the final purification at the end of time. In this way, Constantinople even challenged the apocalyptic importance of Babylonia and Rome.[40]

In the fourteenth and fifteenth centuries, plagues and continuous Ottoman military campaigns further reduced the Byzantine domains, weakened the defenses of the city, and cut Constantinople's ties to its hinterland. By the time Mehmed II conquered the city in 1453, its walls marked the entirety of the domains of the Byzantine Empire and there were only 8,500 people left inside to defend the city.[41]

After the conquest, Mehmed II declared Constantinople his new capital, but he made no drastic changes in the architectural or residential patterns of the city until the 1460s, and did not move his retinue from Edirne until a palace was built for him in Constantinople. In the six years after the conquest, the Hagia Sophia was converted into the Friday Mosque of the city. A palace—partly on the site of the Forum Tauri—and a citadel at the Golden Gate of the Theodosian walls

---

[39] For a discussion on Mehmed II's patronage of geographical works, see especially Brotton, *Trading Territories*, 98–103.

[40] Paul J. Alexander, "The Strength of Empire and Capital as Seen Through Byzantine Eyes," *Speculum* 37, no. 3 (1962): 343–5. Stéphane Yerasimos, *Légende d'empire: La fondation de Constantinople et de Sainte-Sophie dans les traditions turques* (Paris: Maisonneuve, 1990), 183.

[41] Yerasimos, *Constantinople: Istanbul's Historical Heritage*, 6–15.

were constructed. In addition, the first religious establishment outside the city walls—the Ayyub al-Ansari mosque complex—and a commercial center with a port and *bedestan* (marketplace) became part of the city's panorama. These first five projects in the city accommodated, appropriated, and replaced aspects of Byzantine Constantinople; at the same time, they addressed immediate concerns and needs.[42] During this period, the sultan and his Janissaries continued to reside in Edirne. But after 1459, the Ottoman sultan and his household moved to Constantinople, never to leave it again.

Before Constantinople, the Ottomans had set their sights on two capital cities that they perceived as strategic locations for further conquest and temporary seats of government. The Ottoman ruler Orhan Ghazi (r. ca. 1324–62) captured the first Ottoman capital, Bursa, in 1326 after a six-year siege. After the conquest, the Ottomans first settled in the city's castle. New mosque complexes, built by the Ottoman ruler and his military commanders, soon expanded the city beyond its walls. Orhan built a palace for himself in Bursa and declared the city his capital. He also minted the first Ottoman coins there.[43] Until the armies of the Mongol ruler Tamerlane sacked the city in the early fifteenth century, Bursa grew as an important center for the silk and spice trades that passed through the Levant.[44] During this period, the Ottoman state apparatus was still governed informally. Since the Ottomans were still a ghazi frontier state, the members of the ruling family and the family's entourage shared offices.

These practices continued after Edirne became the second Ottoman capital after its conquest in 1369. Edirne's strategic location allowed for further Ottoman expansion in Europe. It was very popular among the members of the dynasty because of its large forests (used for hunting), its beautiful gardens, and its palace. Ottoman rulers, however, were still very mobile, continuing to travel between their fortresses with their retinues. Moreover, they still spent the summers and winters in different places, a practice of nomadic societies.[45] Both Bursa and Edirne were strategic locations for territorial expansion and commercial activities, but they were not the centers of the state. The center was where the sultan and his retinue could be found—in other words, the center was always in motion.

With the conquest of Constantinople, the Ottomans faced the challenge of establishing a permanent seat for both the state and the dynasty for the first time.

---

42 Kafescioğlu, *Constantinopolis/Istanbul*, 16–53.

43 Halil İnalcık, "Bursa," *EI²*, last accessed via Brill Online. University of Pittsburgh. 7 November 2011; Yunus Uğur, "Bursa," in *Encyclopedia of the Ottoman Empire*, ed. Gábor Ágoston and Bruce Masters (New York: Facts on File, 2009), 105–6.

44 For Bursa and its importance in economic networks in the early modern period, see Haim Gerber, *Economy and Society in an Ottoman City: Bursa, 1600–1700* (Jerusalem: Hebrew University, 1988); Halil İnalcık, "Bursa and the Silk Trade," in *An Economic and Social History of the Ottoman Empire, Vol. 1: 1300–1600*, ed. Halil İnalcık and Donald Quataert (Cambridge: Cambridge University Press, 1994), 256–71; Engin Özendeş, *The First Ottoman Capital: Bursa* (Istanbul: Yapı Endüstri Merkezi Yay., 1999).

45 Imber, *The Ottoman Empire*, 143.

Edirne as "the ancient and holy abode of the ghazis" still held a prestigious place among the Ottoman rulers. However, making Constantinople the capital city meant the marginalization of the ghazis in the state apparatus.[46] The architectural projects attempted in Constantinople during the first six years of Ottoman rule in the city illuminated the conflict between the ghazi ideology of the older frontier Ottoman order and that of Mehmed and his new ruling elite, who aimed at establishing a centralized state.[47] In 1459, Mehmed II ordered the construction of a new palace and a mosque complex. He also invited his ruling elite to build religious and charitable institutions around the city.[48] The effort put forth by the sultan and his elite after 1459 was part of a conscious attempt to turn Byzantine Constantinople into Ottoman Constantinople. While these projects all aimed at "beautifying" the city, the differences between the works of the sultan and those of his viziers shed light on the Ottoman state's hierarchical political apparatus. Mehmed II was busy centralizing the state apparatus by regulating the salaries of his administrators and bureaucrats, as well as the promotion process for those seeking government offices.[49] Along with the new construction projects, Mehmed II introduced policies that sought to repopulate Constantinople. He invited and in many cases forced Muslims, Jews, and Christians from all around his empire, as well as former residents, to settle in the city.[50]

The conquest, construction, and settlement policies of Mehmed's reign had a profound influence on Ottoman state ideology. Specifically, they changed the Ottoman understanding of "capital" and subsequently of "state." Constantinople and its status as the New Rome—the capital of the Byzantine and indeed the Roman Empire—supplied the Ottoman ruling elite with a permanent and central location for its bureaucracy, and with an ideal starting point for further territorial expansion to both the east and west. As the Ottomans ceased to be a semi-nomadic frontier society, state ideology changed accordingly. Mehmed II, who formulated his new imperial authority by decree, fashioned a new style of relationship between himself and his ruling elites. This new vision found expression through imperial decrees, new buildings in Constantinople, and the imperial patronage of artists and intellectuals. Mehmed II's aim to establish an empire triggered a tradition of intellectual and scientific patronage in the Ottoman palace. Constantinople eventually became a vibrant cosmopolitan center where artists and intellectuals from both the East and West exchanged ideas and sought the patronage of the sultan.[51] Among these intellectuals was a cohort of geographers who played a special role in the articulation of Mehmed's vision for his empire.

---

[46] Kafadar, *Between Two Worlds*, 148.

[47] Kafescioğlu, *Constantinopolis/Istanbul*, 16–53.

[48] Ibid., 53–143.

[49] Ibid.; Halil İnalcık, *Fatih Devri Üzerinde Tetkikler ve Vesikalar* (Ankara: TTK, 1995).

[50] Halil İnalcık, "The Policy of Mehmed II toward the Greek Population of Istanbul and the Byzantine Buildings of the City," *Dumbarton Oak Papers* 23 (1970): 231–49.

[51] Brotton, *Trading Territories*, 98–9; Kafescioğlu, *Constantinople/Istanbul*, 1–16.

Ptolemy's *Geographia* and Mehmed's Empire

Mehmed II sponsored the production of maps and geographical accounts, and he did so for two distinct but interrelated reasons: to satisfy his intellectual curiosity about the world, and to acquire symbolic and practical assistance in his endeavor to establish a new empire. Mehmed II's library in the Topkapı Palace housed several copies of Ptolemy's *Geographia* in Greek. After the conquest, Mehmed commissioned a Byzantine geographer by the name of George Amirutzes of Trebizond to compile all the regional maps of the *Geographia* into a single world map, and encouraged him and his sons to prepare a new Arabic translation of the work.[52] While George Trapezountios prepared a Latin translation of Ptolemy's *Almagest* for Mehmed II, a Florentine scholar, Francesco Berlinghieri, decided to dedicate a copy of his treatise on the *Geographia* to him. The printing of the work was completed in 1482, a year after Mehmed II died; thus, the treatise entered the Topkapı Palace Library during the reign of Mehmed's son, Bayezid II (r. 1481–1512).[53] The work consisted of the standard Ptolemaic maps, four additional maps of Italy, Spain, France, and Palestine, and a commentary in verse.[54] In 1479, Mehmed II had commissioned an Italian painter, Gentile Bellini, to prepare his portrait in the contemporary European style and to prepare a map of Venice.[55] We know today that Mehmed also possessed a map of the Balkans and personally sketched the walls of Constantinople before the conquest of the city.[56]

Mehmed II appreciated cartography and understood the ways in which maps could be used to support his political ambitions. He was aware of the prestige attached to the ownership of maps and also of the practical value of maps in his future conquests. His library included the classical works of geography and cosmography, as well as new cartographical productions. He actively commissioned maps and even accepted them as gifts from his rivals in the East and West. Al-Istakhri's *Kitab al-Masalik wa al-Mamalik* (*Book of Roads and Kingdoms*) reflects the intellectual cosmopolitanism of Mehmed's reign. Al-Istakhri was a tenth-century Muslim geographer and possibly a member of the "Balkhi school" of geography, which flourished in tenth-century Baghdad under the patronage of the Abbasid court.[57] The *Kitab al-Masalik* was al-Istakhri's only work. This geographical treatise on

---

52 Ahmet Karamustafa, "Military, Administrative, and Scholarly Maps and Plans," in *The History of Cartography, Vol. 2, Bk. 1, Cartography in the Traditional Islamic and South Asian Societies*, ed. J. B. Harley and David Woodward (Chicago: University of Chicago Press, 1992), 210.

53 Francesco Berlinghieri, *Atlas*, TKS. G. I. 84.

54 Brotton, *Trading Territories*, 90.

55 Franz Babinger, "An Italian Map of the Balkans, Presumably Owned by Mehmed II, the Conqueror (1452–53)," *Imago Mundi* 8 (1951): 8–15.

56 Pinto, "Ways of Seeing," 5; Babinger, *Mehmed the Conqueror*, 81.

57 For al-Istakhri, his works, and his connection to the Balkhi school of geography, see Andre Miquel, "al-Istakhri, abu Ishak İbrahim b. Muhammad al-Farisi al-Karkhi," *EI*[2], last accessed via Brill Online. University of Pittsburgh. 7 November 2011; Gerald R. Tibbets, "The Balkhi School of Geographers," in *The History of Cartography, Vol. 2,*

the topography of the Muslim world became very popular among Muslim scholars, and many copies and translations of it circulated in the Muslim world starting in the eleventh century. The work reached Constantinople when it was gifted to Mehmed II by the Aqquyunlu ruler of the time, Uzun Hasan (d. 1478).[58] Hasan rose to power in Iran, Iraq, and Southeastern Anatolia in the fifteenth century, and proceeded to challenge the Ottoman domains. In 1464, he sent an embassy to Venice and offered an anti-Ottman alliance which Venice welcomed. Uzun Hasan openly demonstrated that he was the enemy of the Ottoman sultan. He was, however, aware that Mehmed II valued maps and geographical works. Uzun Hasan took advantage of this knowledge and sent the work to the Ottoman sultan as a gift on the occasion of the latter's succession to the throne. Uzun Hasan was possibly seeking good relations with the Ottoman sultan at the time.

Mehmed's interests were not limited to classical works of geography and cartography. He also actively patronized the production of visual and literary depictions of Constantinople. City views—especially of Venice, Florence, and Jerusalem—became popular toward the end of the fifteenth century throughout Europe.[59] During the same period, bird's-eye views of Constantinople appeared several times in Ptolemy's *Geographia* and in Cristoforo Buondelmonti's *Liber Insularum Archipelagi.* While the images of Constantinople in the copies of the *Insularum* from the 1460s and 1480s emphasized Byzantine monuments and omitted Ottoman buildings,[60] Çiğdem Kafescioğlu has demonstrated that a copy of the *Insularum* produced around 1480 contained an image of Constantinople that underscored both the Byzantine legacy and the Ottoman identity of the city. It is highly plausible that a member of the Ottoman ruling elite—possibly the sultan himself—partook in the production of this image.[61] During the same period, the Ottoman sultan also commissioned literary depictions of Constantinople, which he intended to gift to the rulers of the Muslim world. These works, tailored for Muslim audiences, emphasized images of the conquest and Islamization of the city.[62]

Bk. 1, Cartography in the Traditional Islamic and South Asian Societies, ed. J. B. Harley and David Woodward (Chicago: University of Chicago Press, 1992), 108–10.

58 Pinto, "Ways of Seeing," 52–3. There are nine manuscript copies of al-Istakhri's work in the manuscript libraries of Istanbul. Seven of these copies arrived in Istanbul between the 1470s and 1520. Pinto, "Ways of Seeing," 56; Tibbets, "The Balkhi School," 130–32.

59 Hilary Ballon and David Friedman, "Portraying the City in Early Modern Europe," in *The History of Cartography, Vol. 3, Bk. 1, Cartography in the European Renaissance*, ed. J. B. Harley and David Woodward (Chicago: University of Chicago Press, 2007), 687–90.

60 In a similar fashion, *portolan* charts produced after the Ottoman conquest of Constantinople either kept the Paleologue flag to indicate the city or omitted the flag altogether to deny the Ottoman presence in the city. Campbell, "Portolan Charts," 399–400.

61 Kafescioğlu, *Constantinopolis/Istanbul*, 144–8.

62 Ibid., 170–77. Ian Manners, "Constructing the Image of a City: The Representation of Constantinople in Christopher Buondelmonti's *Liber Insularum Archipelagi*," *Annals of the American Association of Geography* 87 (1997): 72–102.

The acquisition of Ptolemy's and al-Istakhri's works by Mehmed II's library in Constantinople and Mehmed's involvement in the production of the city's images suggest that the Ottoman ruler actively employed visual and literary representations of the world and his new capital city. This was in part due to his intellectual curiosity about the world. But more importantly it was a conscious effort to return Constantinople to its former glory as a cosmopolitan center of intellectual and economic exchange. Through these geographical works, he was also able to shape the image of a new empire, which he created by combining both Byzantine and Muslim traditions.[63] Mehmed was a man of his age; he was a patron of artists and intellectuals. This, combined with the new capital and the new empire, was a legacy he left to his successors.

Bayezid II and Selim I: Constantinople in the Age of Discovery

During the reign of Mehmed's son Bayezid II (r. 1481–1512), international policies as well as building projects in Constantinople aimed to consolidate the cultural legacy of the Muslim world. Sultan Bayezid invested greatly in public works in Istanbul to help emphasize the growing imperial and Islamic character of the city and its inhabitants.[64] He was a pioneer in converting the remaining Byzantine churches to mosques and in building new religious complexes in Constantinople. These efforts coincided with the rise of the Shi'i Safavid dynasty in Iran. As the Safavids adopted the Twelver Shi'i ideology under Ismail, the Ottomans started reformulating their imperial ideology to emphasize Sunni Islam's integral part in their state formation.

Bayezid was aware of the need to bolster his image as the most prestigious and pious Muslim sovereign of his day and to provide cultural legitimacy to the state that his father had founded. The sultan commissioned Idris-i Bidlisi (d. 1520), an experienced chancery official, to write the first comprehensive history of the Ottoman dynasty. Bidlisi composed the work, entitled *Hasht Bihisht* (*Eight Paradises*), in Persian. Bayezid II also requested that he and Kemal Paşazade (d. 1534)—a *madrasa* teacher—write *Tevarih-i Al-i Osman* (*History of the Ottoman House*) in Ottoman Turkish.[65] Both of these historians called attention to the universal rulership of Bayezid II through their emphasis on the centrality of Constantinople in contemporary geographical and political representations. For Bidlisi, Constantinople was the seat of the kingdom and the throne of sovereign rulers. Just like Rome, Bidlisi asserted, Constantinople's seven hills were on the shores of the Roman Sea's green waters.[66] Kemal Paşazade, who emphasized

63 Kafescioğlu, *Constantinopolis/Istanbul*, 154–61.

64 Turan, "The Sultan's Favorite," 158; Doğan Kuban, *Istanbul, Bir Kent Tarihi: Bizantion, Konstantinopolis, Istanbul*, trans. Zeynep Rona (Istanbul: Turkiye Ekonomik ve Toplumsal Tarih Vakfi, 1996), 220–22.

65 Fleischer, *Bureaucrat and Intellectual*, 238–9.

66 Idris-i Bidlisi, *Hasht Bihisht*, Süleymaniye Library, Esad Efendi 2198, 65a, 36a; cf. Kafescioğlu, *Constantinopolis/Istanbul*, 175.

the Islamization of Constantinople immediately after its conquest, compared the city to a universe where flowers are in bloom year round, where drinking houses sparkle like an elegant throne, where mansions are luminous like sun-drenched parlors, and where places of worship shine like the pearl-set crown of Venus. The city, in Kemal Paşazade's poetry, was more beautiful than Kandahar and had a silky complexion like that of Rome.[67]

Bayezid II, who was known to be more religious than his father, did not radically change Mehmed II's tradition of offering patronage to intellectuals from the East and West. He welcomed both Italian and Persian artists and their works to his court. For instance, he tried to commission Leonardo da Vinci and Michelangelo to work on imperial projects in Constantinople.[68] After the death of Mehmed II, Berlinghieri dedicated one of his printed copies of Ptolemy's *Geographia* to Bayezid II and another to Bayezid's brother Cem, who fled to Rhodes after 1482 and was then taken by the Knights Hospitaller to France as a hostage, to be subsequently handed over to the pope.[69] Printed maps proliferated in Europe during this period. A vibrant map trade had begun to flourish in Florence by the beginning of the sixteenth century. While we do not know how widely these printed maps circulated in Constantinople, we do know that Berlinghieri's copy reached the palace library during Bayezid's reign.

Bayezid's son, Selim I, ascended to the throne in 1512. Sultan Selim devoted the first few years of his short reign to dealing with the Safavids. After his conquests of Syria, Egypt, Palestine, and the Hejaz in 1516–17, Selim's empire looked significantly different from his predecessors' in three ways: first, the Ottoman domains now stretched to the Red Sea, giving the Ottomans an opportunity to navigate the Indian Ocean.[70] Second, as a result of Selim I's conquests and Ottoman navy advances, the Ottomans controlled the resources and the maritime routes of the eastern Mediterranean. Finally, Ottoman sultans now ruled over a predominantly Muslim population. Selim's conquests of Mecca, Medina, and Jerusalem provided the Ottomans with legitimacy in the Islamic world and earned them the designation "Servants of the two Sanctuaries," a title that Ottoman sultans used repeatedly.

During his short reign, Selim focused on military campaigns and devoted very little attention to Constantinople. He never implemented any architectural programs and left no new buildings behind. The artists and craftsmen whom he brought from Tabriz and Cairo, however, laid the foundation for future architectural programs

---

[67] Kemal Paşazede, *Tevarih-i Al-i Osman: VII. Defter*, ed. Şerafettin Turan (Ankara: TTK, 1991), 76–7.

[68] Kafescioğlu, *Constantinopolis/Istanbul*, 213.

[69] Brotton, *Trading Territories*, 93–8. On Berlinghieri's efforts to increase the political influence of Florence in Constantinople, see Sean E. Roberts, "Cartography Between Cultures: Francesco Berlinghieri's 'Geographia' of 1482" (PhD diss., University of Michigan, 2006), 1–31.

[70] Casale, *The Ottoman Age of Exploration*, 22–31.

in the city.[71] Piri Reis's world map—which Piri presented to the sultan in 1517 on the eve of Selim's successful campaign in Egypt—and Ali Ekber Khitayi's travel account on China, *Khitay-nameh* (*Book of China*), finished in 1516, are two significant examples of geographical works completed in this period.[72] Although neither of these works focused specifically on Constantinople, they both brought the available geographical knowledge about far-off regions of the world to the center of the Ottoman Empire.

Constantinople in Ottoman Canonical Geography

Süleyman I ascended to the throne in 1520. Early on in his reign, the young sultan successfully seized Belgrade—the key to Hungary—and the Mediterranean island of Rhodes. By doing so, he sent a message to the Muslim and Christian worlds that he was the defender of Islam and a would-be world conqueror. During the first half of Süleyman's reign, as part of his imperial enterprise, his ruling elites redesigned the Ottoman imperial capital of Constantinople and emphasized its role as the center of this new universal empire. The city was now home to the empire's bureaucracy and administrative offices. It was a cosmopolitan metropolis and a bustling hub of commerce, all of which made it—according to Ottoman logic—the center of the world.

The grand vizier İbrahim Pasha was the mastermind behind this new vision and its articulation in artistic and intellectual works. İbrahim Pasha redesigned the imperial ceremonies and the Topkapı Palace to highlight the sultan's military

---

[71] Yerasimos, *Constantinople: Istanbul's Historical Heritage*, 250; Necipoğlu, *Architecture, Ceremonial, and Power*, 22.

[72] On Khitay-nameh and Ali Ekber Khitayi see Kaveh Louis Hemmat, "Children of Cain in the Land of Error: A Central Asian Merchant's Treatise on Government and Society in Ming China," *Comparative Studies of South Asia, Africa, and the Middle East* 30, no. 3 (2010): 434–48; Lin Yih-Min, *Ali Ekber'in Hitayname Adlı* Eserinin Çin Kaynakları ile Mukayese ve Tenkidi (Taipei, 1967); idem, "A Comparative and Critical Study of Ali Ekber's Khitay-nama with Reference to Chinese Sources," *Central Asiatic Journal* 27 (1983): 58–78; P. Kahle, "Eine Islamische Quelle über China um 1500 (Das Khitayname des Ali Ekber)," in *Reprint of Text and Studies on the Historical Geography and Topography of East Asia*, ed. Fuat Sezgin et al. (Frankfurt: IGAIW, 1993), 379–98; Charles Schefer, "Trois Chapitres du Khitay Nameh: Texte Persan et Traduction Française," in ibid., 159–214; J. Th. Zenker, "Das chinesische Reich, nach dem türkischen Khatainame," in ibid., 109–29; H. L. Fleishcer, "Über das türkische Chatai-name," in ibid., 96–107; Ildiko Beller-Hahn, "Ottoman Perception of China," in *Comite International D'etudes Pre-Ottomanes et Ottomanes, VIth Symposium, Cambridge, 1st–4th July 1984: Proceedings*, ed. Jean-Louis Bacque-Grammont and E. J. van Donzel (Istanbul: Divit Press, 1987): 55–64; Fuat Sezgin and Eckhardt Neubauerm, eds., *The Book on China: Khitaynama* (Frankfurt: IGAIW, 1994). Iraj Afshar prepared the first crtitical edition of the Persian text. Ali Ekber Khitayi, *Khitay-nameh: Sharh-i Mushahidat-i Sayyid 'Ali Ekber Khita'i: Mu'asir-i Shah İsma'il Safavid dar Chin*, ed. Iraj Afshar, 2nd ed. (Tehran: Markaz-i Asnad-i Farhang-i Asya, 1993).

victories, his magnificent persona, and the city's centrality in the universe.[73] Under his guidance, most buildings in the second courtyard of the imperial palace—such as the treasury, the tower of justice, and the council hall—as well as the chamber of petitions in the third courtyard, were either rebuilt or remodeled.[74] Together with these restoration activities, palace ceremonial was also changed. While royal processions became rare occasions, the sultan behaved more and more like an idol during his public appearances, especially during the visits of foreign ambassadors. Visitors were not allowed to look at his face, and he remained still and quiet in their presence. The rigid rules of protocol in these ceremonies reflected the structure of Ottoman political order and reminded visitors from both within and outside the Ottoman Empire of the sultan's omnipotence and unparalleled magnificence.[75]

In 1526, a young poet, biographer, and a clerk of the state, Abdü'l-latif (Latifi) (d. 1582), described the Topkapı Palace in his literary treatise, *Risale-i Tarifname-i İstanbul* or *Evsaf-ı İstanbul* (*Qualities of Istanbul*),[76] as follows:

> It is the center of the revolving universe and the circle marked out by the pole that it is located above the two seas. Under the bursting of the seas of the genealogies of its Creator, it is an allusion to "there are rivers under it," or it is a pointer to the inhabited house (heavenly portal type of the Ka'ba) and the waters under the throne of God.[77]

This passage on the Topkapı Palace highlights its elegance, beauty, and location. Its mention of the two seas (the Black Sea and the Mediterranean) alludes to the palace's location at the tip of a historical peninsula. Latifi maintains that the palace is at the center of the universe and likens it to a Qur'anical paradise.[78]

In addition to being the imperial center of the empire, Constantinople also served as the commercial hub for the Mediterranean, welcoming Muslims and

---

73 Necipoğlu, *Architecture, Ceremonial, and Power*, 22–30.

74 Ibid., 76–110.

75 Ibid., 22–30.

76 According to Aşık Çelebi, the famous Ottoman biographer Latifi wrote his *Qualities of Istanbul* during the reign of Süleyman, but presented it to Murad III in 1574 after making minor changes to its introduction. Aşık Çelebi, *Meşa'irü'ş-Şu'ara*, ed. G. M. Meredith-Owens (Cambridge: E. J. W. Gibb Memorial, 1971), 106b–107b. Latifi, in his own words, employed an experimental style in prose to recount the topographical, social, and urban features of Constantinople. Latifi, *Tezkire-i Latifi* (Istanbul: İkdam Matbaası, 1896–97), 65–83.

77 Latifi, *Evsaf-ı İstanbul*, ed. Nermin Suner (Pekin) (Istanbul, Baha Matbaasi, 1977), 16.

78 Qur'an, Surah 9:72: Allah hath promised to believers men and women gardens under which rivers flow to dwell therein and beautiful mansions in gardens of everlasting bliss. But the greatest bliss in the Good Pleasure of Allah: that is the supreme felicity.

non-Muslims alike in the sixteenth century.[79] To preserve and protect commerce in the city, Mehmed II ordered the construction of Constantinople's *bedestan*—the largest covered marketplace in the known world—immediately after the city's conquest.[80] The *bedestan* was part of a larger urban complex, which aimed to coordinate and control the city's commerce and wealth. It was also a space of economic and social interaction for Muslims and non-Muslims.[81] The *bedestan*'s lively and cosmopolitan atmosphere contributed to Constantinople's urban character and enabled officials to consolidate the imperial policies of the Ottoman rulers who used the city as a platform to advocate the centrality of Constantinople in the empire's and world's economy. Moreover, the *bedestan* helped to repopulate the city and revive its economic life—neither the population nor the economy had fully recovered from the Latin conquest of 1204—and to reestablish links with the hinterland that had been cut off during the Ottomans' encirclement of the city.[82]

Soon after its completion, the *bedestan* became the main commercial center of Constantinople and of the eastern Mediterranean. Financial transactions and the trade of valuable imported goods occurred there; it was also used for the storage of money and precious materials. With the wealthiest merchants of the city owning shops there, the market attracted merchants from all over the Mediterranean.[83] According to contemporary endowment deeds and registers, in 1473, there were 124 shops inside the *bedestan* and 72 shops outside of it. The shops included those of silk merchants, cap makers, used-goods merchants, boot makers, saddlers, and cotton and wool dressers. By 1489, there were 126 shops inside the *bedestan*, while the number of shops surrounding it had risen to 782. In 1520, after surviving a fire in 1516, the number of shops in the *bedestan* increased to 168.[84] According to contemporary European accounts, shop owners were Armenians, Jews, Turks, and European merchants.[85] According to registers from 1520, while Muslims occupied 123 of the 168 shops, non-Muslims rented 34 shops; the officials in charge of the *bedestan* occupied 4 corner shops; and 5 corner shops were vacant. Of the non-Muslims, 18 merchants were Jewish, 13 were Armenian, 2 were Greek, and 1 was European. There was 1 female shop owner among the Muslims.[86]

In reference to the cosmopolitan character of the city and its importance in the world economic network, Latifi notes in the opening pages of his treatise:

---

79 Suraiya Faroqhi, *The Ottoman Empire and the World Around It* (New York: I. B. Tauris, 2004), 1–17.

80 Kafescioğlu, *Constantinopolis/Istanbul*, 47–85.

81 Ibid., 73–8.

82 Ibid., 47. Halil İnalcık, "The Hub of the City: The Bedestan of Istanbul," *International Journal of Turkish Studies* 1 (1980): 1–17.

83 Kafescioğlu, *Constantinopolis/Istanbul*, 65–6.

84 İnalcık, "The Hub of the City."

85 Hans Derschwam, *Tagebuch einer Reise nach Konstantinopel und Kleinasien*, ed. Franz Babinger (Munich and Leipzig, 1923), 93.

86 İnalcık, "The Hub of the City."

> It is such a large city ornamented like a paradise. Each side of it is a clime and each district is a sublime city by itself. Places like China, India, Aden and Yemen, which are full of wonders, can be lost in one of its corners [...] Seventy-two different nations live in this city. Each conducts a different type of trade. Its bazaars are much more special than those of India; every corner is a meeting place, like the bazaar of Mina.[87]

Latifi continues to recount the lively atmosphere of trade in the city by describing the goods circulating in the *bedestan* as follows: "What is more, all the marvels from land and sea that attract attention, desire, and esteem, the precious and memorable wonders that are made from silver and stone that arrive there, are valued and esteemed, and are given their due. The buyers and sellers of the world market always gather there."[88] Latifi observes the marketplace of Constantinople with admiration and asserts that there is no other place in the world like it. He compares the city to a chameleon, thanks to its crowded bazaars, lively streets, variety of merchandise, and diversity of people.

Latifi's account allows one to understand how a sixteenth-century Ottoman intellectual perceived and related to the imperial capital. Latifi was a typical Ottoman intellectual of the sixteenth century. Although he was a native of Kastamonu, a small town on the Black Sea coast, he spent most of his life serving as a secretary for public kitchens around the empire. Later, he became an important member of the empire's literary circle. Latifi completed *Tezkire-i Şu'ara*, an acclaimed biographical study of Ottoman literature, in 1546 and presented it to Süleyman I the same year.[89] Süleyman was pleased by the work and appointed him secretary of the Ayyub al-Ansari complex endowment.[90]

Written at a time of political and territorial expansion, Latifi's *Qualities of Istanbul* highlights Constantinople's unrivaled beauty and its centrality in the world, and articulates the Ottoman court's ubiquitous claims to universal sovereignty. Latifi focuses exclusively on the Ottoman imperial capital and tries to capture the nature of the city in both poems and prose. He repeatedly compares the city with other parts of the world, thereby shedding light on the urban and cosmopolitan character of the city. Latifi explains that he found paradise on earth when he

---

87 Latifi, *Evsaf-ı İstanbul*, 9.

88 Ibid.

89 Latifi's *Tezkire* consists of an introduction, three chapters, and a conclusion. In the first chapter, he mentions poets who lived in Anatolia. In the second, he speaks about Ottoman sultans who engaged in poetry. In the third, he mentions, in alphabetical order, the names of 300 poets who lived from the reign of Murad II (1421–51) until 1543. The book is arranged alphabetically, a method that the author introduced to Ottoman biographical works. Nihad M. Çetin, "Latifi," *EI*^{2}, last accessed via Brill Online. University of Pittsburgh. 7 November 2011. For a discussion on Latifi's *Tezkire-i Şu'ara*, see Walter Andrews, "The Tezkere-i Şu'ara of Latifi as a Source for the Critical Evaluation of Ottoman Poetry" (PhD diss., University of Michigan, 1970).

90 Çetin, "Latifi," *EI*^{2}.

came to Constantinople, and that for him, the city was the brightest part of the universe.[91] He even elevates Constantinople above other religious centers, such as Mecca and Medina. In a section dedicated to the Hagia Sophia and Fatih mosques, Latifi assserts that "these two solidly-built and well-arranged places [...] are most marvelous great buildings and stunning foundations without any resemblance. Each of them is a venerable holy place for Muslims where commoners and dignitaries pray alike."[92] After highlighting each building's architectural splendor and spiritual eminence, Latifi concludes, "O you who did not perform his prayer in the city of Istanbul / Don't exclaim that you are a Muslim."[93] In Latifi's words, Constantinople was the political, economic, and religious center of the world: a cosmopolitan center from which imperial power radiated through the Ottoman realm and beyond.

A particular representation of geographic knowledge that shared the same vision as Latifi's distinguished text reached Constantinople in 1525. On the eve of Sultan Süleyman's second campaign to Hungary, an Ottoman geographer by the name of Mustafa b. Ali al-Muwakkit presented the sultan with an account entitled *İ'lamü'l-'İbad fi A'lami'l-Bilad* (*Public Instructions for the Distances of Countries*). As noted in Chapter 1, the work lists the coordinates of 100 cities located in the Northern Hemisphere and provides the distances of 99 of these cities from Constantinople. No world maps or regional charts accompany the text in the *Public Instructions*, but the work is full of details like the length of daytime in each city, the shortest day in hours and minutes, and the degree of the azimuth in each city when facing Mecca. These facts lead one to speculate about the author's intentions in writing this book. At first glance, the work is an instructional treatise possibly written for students and *madrasa* teachers. It is also clear that Mustafa b. Ali is interested in updating the coordinates of the 99 cities in question. He recalculates the coordinates that had been worked out by Muslim geographers and comes very close to the "modern" coordinates.[94] Similarly, in his other well-known work on universal astronomy and geography, *Tuhfetü'z-Zaman and Haridetü'l-Avan* (*The Gift of the Time and the Pearl of the Epoch*), he provides the distances of 150 cities (located in seven climes) from Constantinople.[95]

Mustafa b. Ali opens his *Public Instructions* with a description of Constantinople. He explains at length its geographical location, to which clime

---

91 Latifi, *Evsaf-ı İstanbul*, 7.

92 Ibid., 28.

93 Ibid., 34.

94 Sezgin, *Geschichte des Arabischen Schrifttums, Band X*, 185–91.

95 This work was a translation of mostly Arabic works on cosmography and the wonders of the universe. The third chapter, where the author outlines the distances of 150 cities from Constantinople, is Mustafa b. Ali's own contribution. Franz Taeschner, "Die Geographische Literatur der Osmanen," *ZDMG* 77, no. 2 (1923): 46; Kemal Özdemir, *Ottoman Nautical Charts: The Atlas of Ali Macar Reis*, trans. P. Mary Işın (Istanbul: Marmara Bank Publication, 1992), 28; İhsanoğlu, *Osmanlı Coğrafya Literatürü Tarihi*, vol. 1, 51–2.

it belongs, and the length of daylight and the azimuth of its direction to Mecca. He then proceeds with the urban centers closest to Constantinople. The first cities that Mustafa b. Ali visits are Bursa and Nicea, once the Ottoman capitals. Having replaced the previous centers such as the Ka'ba or Baghdad with Constantinople as the center of the universe, in the rest of his account, Mustafa b. Ali directs most of his attention to the urban centers around the eastern Mediterranean, in particular the geographical regions of the Balkans, Anatolia, the Aegean Sea, and North Africa. While the work covers eight cities from Persia and Iraq, and 10 cities from Central Asia, China, and India, Spain and Italy are represented with 10 cities. The Indian Ocean is almost entirely omitted except for Aden.[96] Mustafa b. Ali's concentration on the eastern Mediterranean stands in direct opposition to the earlier coordinate tables that exclusively focus on the Muslim urban centers.

In this account, Mustafa b. Ali tests and reformulates the results of previous geographical endeavors and places the center of the universe in Constantinople. In doing so, he challenges several Muslim and European geographical traditions. Whereas coordinate tables prepared by Muslim geographers in the pre-Ottoman era frequently took Baghdad as the center, Mustafa b. Ali places Constantinople at the core and provides the distances of each city from the Ottoman capital. Likewise, whereas the authors of classical Islamic pilgrimage literature took Mecca as their reference point when noting distances, Mustafa b. Ali prefers to calculate distances from Constantinople. *Public Instructions* also responds to the medieval and early modern European geographical discourse that centered the world around Jerusalem or Rome. In this sense, Mustafa b. Ali's work represents a stark departure from both Muslim and European geographical traditions and epitomizes a geographical consciousness specific to the Ottoman court in the service of its imperial ideology.

Mustafa b. Ali dedicated the work to the sultan. However, he employed very technical language and offered meticulous geographical details about the 100 cities. It would be safe to assume that the author tried to impress the sultan and

---

[96] Mustafa b. Ali provides the distances of 99 locations from Constantinople in the following order: Bursa, Iznik, Kütahya, Afyonkarahisar, Konya, Aksaray, Kayseri, Ankara, Kastamonu, Sinop, Amasya, Samsun, Sivas, Tarsus, Aleppo, Tripoli (Lebanon), Damascus, Gaza, Jerusalem, Cyprus, Antalya, Manisa, Rhodes, Damietta, Rosetta, Alexandria, Cairo, Medina, Mecca, Saba' Island, Sana'a, Aden, Sohar, Abyssinia, Equator, Trabzon, Erzincan, Diyarbakır, Mosul, Baghdad, Tabriz, Shirvan, Hamadan, Isfahan, Shiraz, Hormuz, Bukhara, Samarqand, Derbent, Tblisi, Semendur, Feodosiya, Bolgary, Tobolsk, Thule, Uzgen, Khotyn, Karakorum, El Mansoura, Multan, Kandahar, Kannauj, Sri Lanka, Köfü (I was not able to determine the present-day location of this place. The information on the original text does not provide any clues beyond its distance from Constantinople and the fact that this location is off the coast of China [*Köfü min-es Sin*]), Hangzhou, city of Gog, city of Magog, Çorlu, Edirne, Malkara,Gallipoli, Plovdiv, Sofia, Smederevo, Akkerman, Sarajevo, Skopje, Saros, Thessalonica, Lesbos, Chios, Athens, Euboea, Vlorë, Brindisi, Neapoli, Rome, Venice, Piombino, Genoa, Tripoli (Libya), Mahdia, Tunis, Cordoba, Marrakesh, Toledo, Fez, Tangier, Sousse.

contemporary ruling elites and intellectuals with his geographical knowledge and astronomical analysis. Mustafa b. Ali's *Public Instructions* remained popular for at least a century after his death and inspired numerous manuscript copies even after the sixteenth century. Thirty-seven copies of the work survive today. Among these copies, 6 date to the sixteenth century, 11 to the seventeenth, 12 to the eighteenth, and 8 to the nineteenth.[97]

Circumstantial evidence suggests that Mustafa b. Ali did indeed impress the Ottoman sultan. When the geographer died in 1571, he was the *müneccimbaşı*, or chief court astrologer. In the Ottoman palace structure, this position was considered very prestigious and influential. The candidates were chosen from among the *ulama*, the religious hierarchy. *Müneccimbaşıs* were responsible for preparing calendars, choosing the perfect time to launch a military campaign, calculating the birth date of future sultans, and determining the time of the imperial weddings.[98] Either Süleyman or his son Selim II can be credited with appointing Mustafa b. Ali *müneccimbaşı* to the palace.

Close contextual analysis of *Public Instructions* allows one to visualize the connection between developing geographical consciousness and the imperial enterprise in the sixteenth century Ottoman Empire. Mustafa b. Ali wrote at the time when Süleyman I led campaigns in the Mediterranean and in Southeastern Europe. In the wake of the conquests of Belgrade (1521) and Rhodes (1522), Süleyman became a serious adversary of rulers on both sides of the Mediterranean: Charles V of the Holy Roman Empire and the Safavid shah, Tahmasb I (r. 1524–76), who actively tried to stop the Ottomans in the Mediterranean and to drive them out of Hungary and the Indian Ocean respectively.[99] Mustafa b. Ali's work lent support to Ottoman campaigns by confirming the Ottoman Empire's centrality in the universe and therefore justifying the Ottomans' attempts to expand their territory.

It would be wrong to assume that the Ottomans were alone in making claims for centrality. In the pre-Ottoman Muslim world, the capital cities of the caliphates—Damascus, Cairo, and Baghdad—held important roles as administrative and cultural centers. Most important of all was the city of Mecca (where the Ka'ba is located), considered the religious center of the Muslim world. In Europe, the Habsburg monarchy had used Vienna as the family's residential seat since the fifteenth century. It was not, however, until the second half of the sixteenth century that the Habsburg Empire built an administrative center. During the reign of Charles V, the monarchy had several administrative centers—including Madrid, Rome, and Brussels—as well as imperial cities, such as Nuremberg, Ulm, and Augsburg. Charles had to travel back and forth between these cities to run his empire.[100] Only in 1561 did his son Philip II decide to establish a fixed center

[97] İhsanoğlu, *Osmanlı Coğrafya Literatürü Tarihi*, vol. 1, 49–51.

[98] İsmail Hakkı Uzunçarşılı, *Osmanlı Devletinin Saray Teşkilatı* (Ankara: TTK, 1988), 369–70.

[99] In the same period, the Portuguese also sought an anti-Ottoman alliance with Shah Tahmasb I. See Casale, *The Ottoman Age of Exploration*, 50.

[100] Kleinschmidt, *Charles V*, 142.

for his government in Madrid. Philip chose the city for its central location, its excellent water supply, and the empty spaces around it, which would allow for further expansion.[101]

In the second half of the sixteenth century, Philip II engaged in a geographical enterprise to update and extend geographical knowledge about his possessions. Around 1560, he launched three cartographical projects in Spain. First, he commissioned the Dutch artist Anton van den Wyngaerde to prepare city views of around 50 Spanish cities. The second project was the creation of a complete map of the Iberian Peninsula, carried out by the empire's principal cartographer, Pedro de Esquivel, in the 1570s and 1580s. A team of cartographers led by de Esquivel surveyed the entire Iberian Peninsula and compiled the results in an atlas of 21 pages. The last geographic project initiated by Philip consisted of a series of questionnaires about the economic resources of the empire; the questionnaires were sent out to various communities in the 1570s. None of these projects was completed. Nevertheless, Philip's effort to gather such an immense body of knowledge about his domain was unprecedented.[102] In commissioning both geographic and chorographic views of Spain, Philip was emphasizing two ideologies about the nature of his empire: Wyngaerde's paintings of Spanish cities presented Spain as a group of realms or city-states, while the Escorial Atlas depicted Spain as a universal empire ruled by a king in Madrid with the help of his centralized bureaucracy.[103]

In the Ottoman Empire, a general mapping of the domains was never attempted. Although the court encouraged individual geographers like Mustafa b. Ali to prepare geographical accounts of the Ottoman domains, it did not attempt to organize or standardize the content of their works. One plausible explanation for this lack of interest in geographical works was the systematic land/revenue surveys (*tahrir*) that the Ottoman state had at its disposal.[104] These surveys mapped the economic conditions, size, and composition of the empire's population.[105] In addition to land surveys, a wide variety of data—including but not limited to provincial law codes, the registers of bestowals of tax benefices (*timar*), and a host of financial records—were available for consultation. The important registers traveled with the army, so that this information was accessible even in wartime.[106]

---

[101] Parker, *The Grand Strategy of Philip II*, 17.

[102] Ibid., 59–65.

[103] Barbara E. Mundy, *The Mapping of New Spain: Indigenous Cartography and the Maps of the Relaciones Geograficas* (Chicago: University of Chicago Press, 1996), 1–11.

[104] On *tahrir* registers, see Heath W. Lowry, "The Ottoman Tahrir Defterleri as a Source for Social and Economic History: Pitfalls and Limitations," in *Studies in Defterology: Ottoman Society in the Fifteenth and Sixteenth Centuries*, ed. Heath W. Lowry (Istanbul: ISIS, 1992), 3–18; Colin Heywood, "Between Historical Myth and Mythohistory: The Limits of Ottoman History," *Byzantine and Modern Greek Studies* 12 (1998): 315–45; Halil İnalcık and Donald Quataert, eds., *An Economic and Social History of the Ottoman Empire, 1300–1914* (Cambridge: Cambridge University Press, 1994).

[105] Ágoston, "Information, Ideology, and Limits of Imperial Power," 80–81.

[106] Ibid., 81.

Also present during military campaigns were the chroniclers who wrote accounts of the victorious campaigns for future generations. Matrakçı Nasuh (d. 1564) was one such chronicler during the reign of Süleyman the Magnificent. Matrakçı was in the retinue of Süleyman when the sultan launched his first imperial campaign in the East against the Safavids. The campaign lasted from 1533 to 1536, when the Ottoman army entered Tabriz in 1534 and Baghdad in 1535. Matrakçı Nasuh composed *Beyan-ı Menazil*, a diary of the campaign, during the army's actual journey; later, he illustrated it with 130 miniature depictions of the major stops en route.

Matrakçı Nasuh's work is one of the few manuscripts that included both visual and narrative representations of major Ottoman urban centers from Constantinople to Baghdad. The *Beyan-ı Menazil* opens with a short introduction in which the author lays out his reasons for composing this work and describes the seven climes of the known world. Here, he refers to Süleyman I as the sultan of the seven climes, a title once reserved for Alexander the Great. After recounting the seven climes, the cities that belong to each of these climes, and the stars that correspond to them, Matrakçı Nasuh continues his narrative with a double-page miniature bird's-eye view of Constantinople (Plate 2), depicting *intra muros* on the right and the Galata region on the left.[107]

In the image of *intra muros*, one can easily discern Hagia Sophia, the Sultan Bayezid II Mosque, the Old Palace, and the Fatih Mosque on one axis in the middle of the picture. In addition to these buildings, Matrakçı Nasuh included illustrations of the *bedestan*, the Topkapı Palace, and the Yedikule dungeons. Between the Galata region and *intra muros*, Ottoman galleys reveal that there was an imperial arsenal in Tophane, while also conveying to the audience Constantinople's status as a port city. Matrakçı does not focus on the city's residential buildings or streets, but rather on imperial mosques, palaces, and other urban monuments.

Matrakçı Nasuh's visual depiction of Constantinople serves as a passage in the work between the introduction, where he describes the seven climes of the world, and the visual and narrative representations of major cities in the Ottoman realm between Istanbul and Baghdad. Most of the cities are featured on their own page and look like small replicas of Istanbul. The main mosque of each city rises up from the city center, while the other urban facilities, such as bazaars and water fountains, surround the mosque. After Istanbul, Tabriz is the second city to which Nasuh dedicates two full pages. Here, the mosque is shown rising from the center, while deer and rabbits roam around the city walls.[108]

The text around the miniature image of Tabriz celebrates the arrival of the Ottoman sultan to the city and introduces one of the main themes of the work—the Ottoman sultan as the dispenser of justice throughout the Ottoman realm:

---

[107] Matrakçı Nasuh, *Beyan-ı Menazil-i Sefer-i 'Iraqeyn-i Sultan Süleyman Han*, Istanbul University, TY. 5964, 8b–9a.

[108] Ibid., 26a.

> After the victorious troops who were in that region [Tabriz] set their sights on reaching eternal success with reverence and honor and respect and consideration, with the appearance of the sultanic and imperial throne and the starlike arrival of the Rumis, they held the court of justice because the celestial-sphere-adorning court of the Sulaiman of the time was necessary for that land.[109]

Matrakçı Nasuh reminds his readers on several occasions that the sultan brought not only security but also justice to the Ottoman realm. Whenever he reached an important station along his campaign route, the sultan, readers are told, set up his *divan-ı adalet* (court of justice). In several miniatures, Matrakçı even depicts the sultan's tent—the physical symbol of his justice and the place where the imperial council convened—surrounded by smaller tents belonging to his retinue.[110]

Looking at this work, a question presents itself: what did Matrakçı Nasuh hope to achieve by composing a detailed account of the imperial tent and the way in which the Ottoman sultan practiced justice? Matrakçı Nasuh, a protégé of the grand vizier Rüstem Pasha, was clearly aiming to consolidate the political agenda of his patron by articulating a vision in which justice held the central place. When Süleyman ascended to the throne, he stated, in accordance with the apocalyptic expectations of the age, that he would disseminate justice around the world as the messianic ruler.[111] By the time Matrakçı Nasuh completed and presented his work to Sultan Süleyman in 1537, the sultan had already executed İbrahim Pasha. In the second half of his reign, the sultan delegated most of his power to imperial bureaucrats and withdrew to his own quarters in the palace. During this period, both the chief jurisprudent and members of the bureaucratic elite systematically codified, compiled, and reconciled dynastic law with Islamic law. As Süleyman came to exude a "somber imperial image," he employed a far less personal icon of rulership: imperial law. Rüstem Pasha, as one of the prominent political figures of this period, clearly influenced Matrakçı's narrative and depictions. Matrakçı Nasuh gave body to this imperial image by portraying the Ottoman sultan as the Law Giver and by reminding his subjects of his role in spreading justice throughout the world. He used both visual and textual mediums to encourage the sultan to spread his authority and justice throughout his realm, and to remind him of the centrality of his city.

Mustafa b. Ali, Latifi, and Matrakçı Nasuh employed both novel and traditional ways of narrating their first-hand observations and to articulate the political agenda of their patrons. Their accounts promoted Constantinople as the center of the universe. They all lived and experienced Constantinople during the tenure of İbrahim Pasha and his imperial vision, which portrayed the Ottoman sultans as the universal sovereigns and world conquerors. Not every Ottoman intellectual, however, supported their claims.

---

109 Ibid., 28b–29a.

110 See, for example, ibid., 13b–14a, 15b, 20b, 24b, 25a, 26b, 29b, 31a, 72a, 73a, 74a, 74b.

111 Fleischer, "The Lawgiver as Messiah," 159–65.

In 1598, the intellectual, traveler, and geographer Aşık Mehmed compiled his voluminous universal cosmography, *Views of the World.* Although the author claims that he compiled the work for posterity and to refresh the memories of his travels,[112] he opens his account with eloquent praise for the sultan of the time, Mehmed III (r. 1595–1603).[113] Clearly written to please the Ottoman sultan, the work is divided into two parts. In the first part, Aşık Mehmed, in a manner reminiscent of the traditional Muslim accounts of universal cosmography, discusses the creation from a religious and philosophical point of view. Here, he describes the sun, moon, angles, djinns, heaven, and hell. The second part, which is further divided into 18 subsections, is where the author delves into the geography of the world. Here, in separate chapters, the author describes the seas, islands, lakes, animals, rivers, hot springs, mountains, cities, treasures, botany, animals, and finally the anatomy of the human body. Throughout the second part, Aşık Mehmed benefits extensively from the traditional accounts of geography.[114]

In the chapter on Constantinople, which the author locates in the sixth clime, Aşık Mehmed employs a variety of pre-Ottoman geographical literature such as Ibn Khordadbeh's *Kitab al-Masalik wa al-Mamalik* (885/886) and al-Qazwini's *Asar al-Bilad wa Ahbar al-'Ibad* (1275/76), Abu al-Fida's *Taqwim al-Buldan* (1321), Ibn al-Wardi's *Haridat al-'Aja'ib* (1419), as well as contemporary Ottoman chronicles such as Hoca Sa'deddin's (d. 1599) *Tacü't-Tevarih* and Ramazanzade's (d. 1571) *Tarih-i Nişancı Mehmed Paşa.*[115] In this chapter, Aşık Mehmed first narrates the story of the foundation of Constantinople. Thereafter, he delves into the history of the conquest and remaking of Constantinople as the Ottoman capital. His lengthy description of the construction of Hagia Sophia, Ayub al-Ansari complex and the Fatih complex, as well as several other mosques around the city that were built by the ruling elites, lacks the vividness and sincerity of Latifi's descriptions of the imperial capital. Aşık Mehmed barely mentions his own observations and mainly reiterates what previous historians and geographers narrated about Constantinople. Furthermore, unlike Mustafa b. Ali, Aşık Mehmed maintains that Mecca is the center of the universe. Aşık Mehmed opens the second part of his work with a description of the Ka'ba and asserts that the "Ka'ba is circumscribed by Masjid al-Haram which is in turn encircled by Mecca. Around Mecca is sacred and beyond the borders of this sacred place is the rest of the world."[116] Aşık Mehmed supports this description with a diagram of the world where he places 11 regions of the world around Mecca (Figure 2.5).

Aşık Mehmed wrote his account almost a decade after Takiyüddin's observatory in Galata was destroyed, as advised by the *Şeyhülislam.* As a recipient of Mehmed III's patronage, Aşık Mehmed probably witnessed the rising influence of the religious hierarchy in the affairs of the imperial court. During this

112 Aşık Mehmed, *Menazirü'l-Avalim*, ed. Mahmut Ak (Ankara: TTK, 2007), vol. 2, 9.

113 Ibid., 5.

114 On the sources of Aşık Mehmed, see ibid., vol. 1, civ–clxiii.

115 On Aşık Mehmed, his career and travels, see ibid., vol. 1, xxxvii–lxix.

116 Ibid., vol. 2, 130.

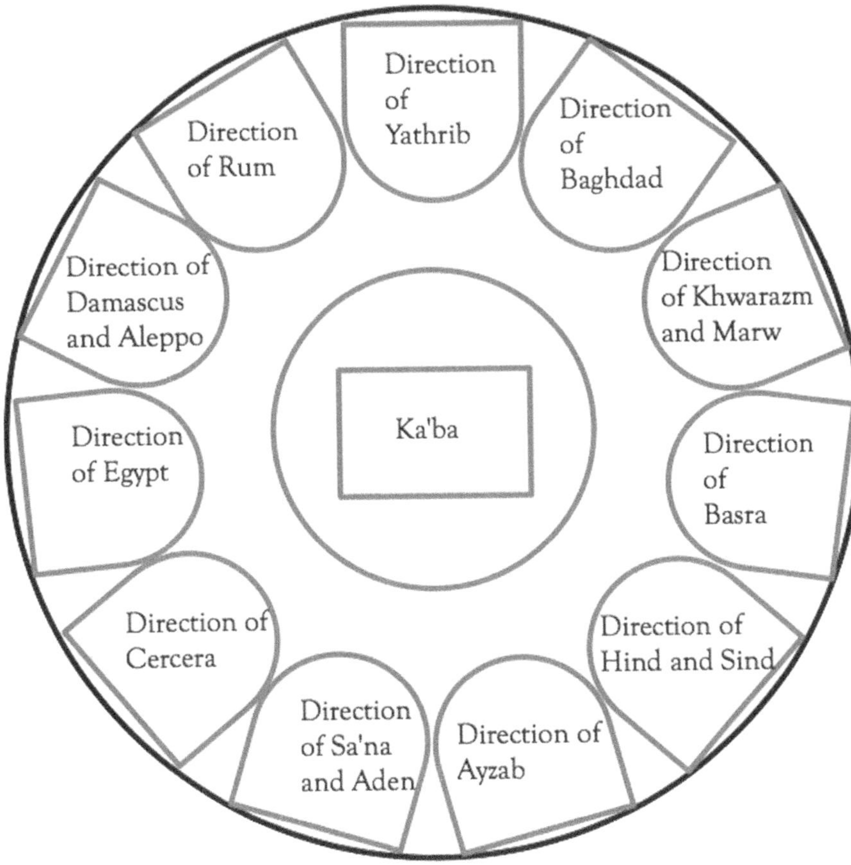

Fig. 2.5 The world according to Aşık Mehmed. Modeled after Aşık Mehmed, *Menazirü'l-Avalim*. Süleymaniye Library, Esad Efendi 2421, 37b.

Note: The Süleymaniye Library, Esad Efendi ms. dates from 1759. It is based on the autograph manuscript and is the closest copy to the original ms. See ibid., vol. 1, xci–xcii.

period, as the provinces were in constant turmoil, the court was under constant fire due to accusations of corruption, and its absolutist tendencies came under a lot of criticism.[117] In order to deal with the crisis, the court asked for help from the religious hierarchy, in particular the *Şeyhülislam*, to re-establish social order. As a result, conservative regulations regarding the clothing of non-Muslims and alcohol consumption reached unprecedented levels.[118] Written during this period, the *Views of the World* was a nod to the traditional accounts of geography and

[117] Tezcan, *The Second Ottoman Empire*, 64–5.

[118] Ibid., 66.

clearly contradicted Mustafa b. Ali's and Latifi's assertions about the importance of Constantinople. Aşık Mehmed was writing at a different time for a different audience. Although the *Views of the World* almost went unread in the sixteenth century,[119] Aşık Mehmed's voluminous work on the geography of the universe remained a legacy for future times.

This chapter has portrayed the changing position of Constantinople in the eyes of the Ottoman court through the lens of the geographical works from the reign of Mehmed II to Mehmed III. Through the analysis of these works, this chapter maintained that after the conquest of Constantinople, the Ottoman court inherited the Byzantine tradition of creating an administrative center that reflected the empire's religious, social, economic, and political pretensions. Before Constantinople, the Ottomans had Bursa and later Edirne as their capital cities. Neither city, however, was considered an administrative or bureaucratic center. Mehmed II's conquest of Constantinople provided Ottoman ruling elites with the elements needed to establish a centralized and bureaucratic empire: the legacy of Byzantium and a strategic location. In the sixteenth century, Ottoman ruling elites took this legacy one step further and portrayed their sultans as world conquerors and universal rulers. A cohort of geographers, especially during the reign of Süleyman the Magnificent, articulated this imperial vision, provided the symbolic support and presented the available geographical knowledge to buttress these claims of the Ottoman court. They formulated and reformulated the importance of the city for the political agendas of their patrons. These geographers were not officially sponsored by the court; however, they prepared their works for the members of the court and sought the patronage of Ottoman sultans. With the help of their works, Ottoman rulers and policymakers in Constantinople redesigned the city to highlight its Ottoman identity, cosmopolitan character, and central location in the world. Constantinople was the center of the empire and the universe. It was the city where the Ottomans' imperial splendor originated and spread throughout the empire and beyond.

119 Among 17 copies of the manuscript, there is only one copy from the sixteenth century. It is the autograph copy of the work. There are seven from the seventeenth century and three from the eighteenth century. The remaining six copies are not dated. Aşık Mehmed, *Menazirü'l-Avalim*, ed. Mahmut Ak, vol. 1, xci–xcv.

Chapter 3
Charting the Mediterranean: The Ottoman Grand Strategy

> With colorful flags and banners, the ships of the fleet became a rose garden ornamented with flowers. White sails covered the face of the blue sky. The sea became a jungle of thickets. The White Sea [Mediterranean] looked like a grey sea with fast and endless number of freshly caulked, black coated, new ships, covered with immeasurable tar to cure and remedy the possibility of rebellion, calamity, and fear.[1]

Sixteenth-century Ottoman historian Kemal Paşazade's vivid description of the Ottoman fleet and the Mediterranean on the eve of the conquest of Rhodes is exemplary of the Ottoman perceptions of the Mediterranean. This sea, which surrounded the Ottoman domains, carried unquestionable importance for the Ottoman imperial claims. For the sixteenth century Ottoman court, the Mediterranean was an object of territorial expansion and constant political and economic attention. It was also a valuable canvas onto which the Ottoman intellectuals and policymakers painted the magnificence and universal sovereignty of their sultans and their empire. In this enterprise, almost as important as the efforts of the activities of the Ottoman fleet, were the cartographical accounts on the Mediterranean, prepared for and commissioned by the Ottoman court.

In 1495, the Ottoman sultan Bayezid II summoned Kemal Reis, a former pirate, to join the Ottoman fleet in its struggle against the Venetians in the eastern Mediterranean. Soon thereafter, Kemal Reis and his nephew Piri Reis (d. 1554) joined forces with Ottoman captains in an effort to help Ottoman ships establish a formidable presence in the Aegean Sea and around the North African coast.[2] Around 1510, Piri Reis withdrew to Gallipoli, where he completed a world map and assembled the notes for his navigational guidebook.[3] He presented the world

---

1 Kemal Paşazade, *Tevarih-i Al-i Osman*, ed. Şefaettin Severcan (Ankara: TTK, 1996), bk. 10, 135.

2 Svat Soucek, "Piri Reis," *EI²*.

3 This paragraph is based on Adnan A. Adıvar, *Osmanlı Türklerinde İlim*, 74–8; A. Afetinan, *Life and Works of Piri Reis*, trans. Leman Yolaç and Engin Uzmen (Ankara: Türk Tarih Kurumu, 1987); Paul Kahle, "Piri Re'is: The Turkish Sailor and Cartographer," *Journal of the Pakistan Historical Society* 4 (1956): 99–108; Klaus Kreiser, "Piri Reis," in *Lexicon zur Geschichte der Kartographie*, ed. Ingrid Kretschmer, Johannes Dörflinger, and Franz Wawrik (Vienna: Franz Deuticke, 1986), vol. 2, 607–9; Soucek, "Piri Reis," *EI²*; idem, "Islamic Charting in the Mediterranean," 267; idem, "Piri Reis," in *Süleyman the Second and His Time*, ed. Halil İnalcık and Cemal Kafadar (Istanbul: ISIS Press, 1993), 379–89; Franz Babinger, "Piri Muhyi'd-Din Re'is," *EI¹*; Fuad Ezgü, "Piri Reis," in *İslam Ansiklopedisi* (Istanbul: MEB, 1964), vol. 9, 561–3.

map to Selim I when the Ottoman sultan was in Cairo in 1517. Selim rewarded the intrepid seaman by assigning him to the imperial sea captains corps in Alexandria.[4] In 1524, Piri Reis led the Ottoman fleet to Egypt to carry out the grand vizier İbrahim Pasha's campaign to suppress Governor Ahmed Pasha's rebellion. During this venture, the grand vizier first saw Piri Reis's guidebook, which the captain had completed in 1521, and commissioned the sea captain to prepare an expanded version of the work for Sultan Süleyman.[5] Piri Reis completed his revisions and presented the final product, *Kitab-ı Bahriye* (*Book of Sea Lore*), to Süleyman I in Istanbul in 1526.

Piri Reis's cartographic works epitomize the developments that occurred in the field of cartography in the sixteenth-century Ottoman Empire. Piri Reis's work as a seaman and as a cartographer points to the informal, experiential, and practical nature of early Ottoman mapmaking. Further, his rapid promotion in the Ottoman military testifies to the rising awareness of geography among members of the Ottoman court. While Piri Reis's contributions enhanced the Ottoman Empire's collection of geographical sources, his most enduring legacy is that he was one of the first cartographers to use Ottoman Turkish in his works. Moreover, that the guidebook he presented to Süleyman in 1526 met with acclaim suggests a significant perceptual shift within the sultan's court: clearly, the Mediterranean was becoming increasingly important to the Ottomans and their goal of becoming universal rulers.

From the viewpoint of political ideology, taking control of the Mediterranean was a milestone in the Ottoman court's efforts to be recognized as universal sovereigns, just like the Roman emperors once were. The Ottomans were not, however, the only imperial power to make these claims. At the other end of the Mediterranean, the Spanish Habsburgs challenged the Ottoman ruling elite's political agenda and imperial vision. This chapter addresses how the Ottoman-Habsburg political and military rivalry over the Mediterranean influenced the production of geographical knowledge in Istanbul and Madrid in the sixteenth century. It argues that the protracted nature of Ottoman-Habsburg military conflicts in the Mediterranean as part of the evolving rivalry between the two empires increased each state's involvement in the production and distribution of cartographical works. The Habsburg-Ottoman political and ideological rivalry helped the growth of cartography as a discipline and as a tool for validating imperial worldviews. In the early modern Mediterranean, itineraries or narrative descriptions of routes were still the main guides for travelers, but as the mapmaking industry evolved in the sixteenth century, maps and charts became useful to rulers, commanders, pilgrims, and merchants. In Istanbul, the awareness of what the Spanish were accomplishing gave the members of the Ottoman court an incentive

---

4 Cf. Casale, *The Ottoman Age of Exploration*, 214; Cengiz Orhonlu, "Hint Kaptanlığı ve Piri Reis," *Belleten* 34, no. 134 (1967): 235–45.

5 Piri Reis, *Kitab-ı Bahriye*, Süleymaniye Library, Ayasofya 2612, 3a. Piri Reis, *Kitab-ı Bahriye*, ed. E. Zekai Ökte, Vahit Çabuk, Tülay Duran, and Robert Bragner (Ankara: Ministry of Culture and Tourism of the Turkish Republic, 1988), vol. 1, 42–3.

to collect or commission maps. Similar to the developments in Europe, maps also gained symbolic importance for the Ottoman patrons who often commissioned them to gain prestige and formulate their own political agenda.

The Spanish Habsburgs and Official Cartography

In the sixteenth century, the Ottomans and the Habsburgs tried to consolidate their empires and exert more control over their domains. Their prevailing model of empire was still the Roman Empire. Without holding the Mediterranean, which Braudel labeled "the center of the world," universal imperial power could not be imagined or convincingly projected.[6] Over the course of the sixteenth century, the Ottomans and the Spanish Habsburgs helped define each other's diplomatic, economic, and religious policies. Both empires shared common imperial, political, and intellectual conceptions of power, all centered on the geographical and ideological importance of the Mediterranean.[7]

The Mediterranean connected both empires to the outer world. It was their link to other cultures and lands. At the same time, it was a unified space that could be charted and measured in precise terms. It was the place where the Ottomans' and the Habsburgs' competing imperial aspirations had to be proven. During this struggle, the imperial courts in both Spain and the Ottoman Empire used cartography—among other military and political strategies—to familiarize themselves with this space and to assert their positions within it. Despite their differing fortunes, Ottoman and Habsburg rulers shared one key interest: maps. Both relied increasingly on cartographic works, which were of military and political importance and, even more importantly, provided the visual representation of their progress toward universal rule.

---

[6] Fernand Braudel, *The Mediterranean and the Mediterranean World in the Age of Philip II*, trans. Sian Reynolds, 2 vols. (New York: Harper and Row, 1972). For the Mediterranean's geographical and conceptual significance in world history, see David Abulafia, "What Is the Mediterranean?" in *The Mediterranean in History*, ed. David Abulafia (London: Thames & Hudson, 2003), 11–32; David Abulafia, "Mediterraneans," in *Rethinking the Mediterranean*, ed. William V. Harris (Oxford: Oxford University Press, 2005), 64–93.

[7] On the religious, cultural, and economic exchanges in the early modern Mediterranean, see Molly Greene, *A Shared World: Christians and Muslims in the Early Modern Mediterranean* (Princeton: Princeton University Press, 2000). On the Mediterranean in the early modern international maritime order, see Molly Greene, *Catholic Pirates and Greek Merchants: A Maritime History of the Early Modern Mediterranean* (Princeton: Princeton University Press, 2010). On the Mediterranean's role in identity formation, see Eric Dursteler, *Venetians in Constantinople: Nation, Identity, and Coexistence in the Early Modern Mediterranean* (Baltimore: Johns Hopkins University Press, 2006); Natalie Rothman, "Becoming Venetian: Conversion and Transformation in the Seventeenth-Century Mediterranean," *Mediterranean Historical Review* 21, no. 1 (2006): 39–75; Natalie Rothman, *Brokering Empire: Trans-Imperial Subjects Between Venice and Istanbul* (Ithaca: Cornell University Press, 2011).

Charles V was surrounded by nobles and advisers who used and collected maps. Among his closest companions were Alonso de Santa Cruz (d. 1567), the empire's official historiographer and royal cosmographer, and the engineer Juanelo Turriano (d. 1585).[8] Charles's first "imperial geographer" was Jacob van Deventer, who also served as the "royal geographer" under Philip II.[9] Official cartography in the Habsburg monarchy, under both Charles and Philip, played a crucial role within the administration and bureaucracy, and in the propagation of the Habsburgs' imperial agenda. While seamen routinely employed maps and itineraries to find their way around the Mediterranean, maps drawn by official cartographers decorated palaces and articulated imperial claims. One particularly striking example was a series of tapestries commemorating Charles V's 1535 conquest of Tunis. The tapestries, which Willem de Pannemaker prepared in Brussels between 1548 and 1554, were displayed during important imperial processions, royal baptisms, marriages, and funerals. The first tapestry shows the Mediterranean in the center, with North Africa at the top and Europe at the bottom. The inscriptions at the top and the bottom celebrate Charles's victory in Tunis over Hayreddin Barbarossa, while the cartographer on the right holding a cartouche makes clear that the designer was guided by contemporary cartographic knowledge when creating this tapestry.[10] The remaining panels provide panoramic views of Barcelona and Valletta (located on the island of Malta), cities central to the Habsburg triumph. For Charles and his audience, the maps advertised the extent of the emperor's rule, which now stretched not only to the Americas, Asia, and the Mediterranean, but also to Africa.[11]

---

[8] David Buisseret, "Spanish Peninsular Cartography, 1500–1700," in *The History of Cartography, Vol. 3, Bk. 1, Cartography in the European Renaissance*, ed. J. B. Harley and David Woodward (Chicago: University of Chicago Press, 2007), 1082.

[9] Coincidentally, this was also the period when Henri II (r. 1547–59) of France established the office of *geographe du roi*. The first person to hold this office was Andre Thevet, a celebrated cosmographer. The office subsequently went to Nicholas de Nicolay who was commissioned to work on a detailed description of the kingdom. Richard Kagan and Benjamin Schmidt, "Maps and the Early Modern State: Official Cartography," in *The History of Cartography, Vol. 3, Bk. 1, Cartography in the European Renaissance*, ed. J. B. Harley and David Woodward (Chicago: University of Chicago Press, 2007), 667. In 1551, Nicholas de Nicolay also visited the Ottoman lands in the entourage of the French ambassador to chronicle the journey and record his observations about the Ottoman court and people. In 1567, he published his *Les qvatre premiers livres des navigations et peregrinations orientales*. On Nicolas de Nicolay and his travel account, see Stephane Yerasimos, *Les Voyageurs Dans L'Empire Ottoman (XIV*e*–XVI*e *siècles): Bibliographie, Itinéraires et Inventaire des Lieux Habités* (Ankara: TTK, 1991), 224–5; Marcus Keller, "Nicolas de Nicolay's *Navigations* and the Domestic Politics of Travel Writing," *L'esprit créateur* 48, no. 1 (2008): 18–31; David Brafman, "Facing West: The Western View of Islam in Nicholas de Nicolay's 'Travels in Turkey,'" *Getty Research Journal* 1 (2009): 153–60.

[10] Lisa Jardine and Jerry Brotton, *Global Interests: Renaissance Art between East and West* (Ithaca: Cornell University Press, 2000), 84.

[11] Kagan and Schmidt, "Maps and the Early Modern State," 671.

The Spanish Habsburgs experienced a transformation in the sixteenth century that made their empire a dominant player in Europe, the Mediterranean, and the Atlantic. During this period, under the reign of Charles V, Spain witnessed its most extensive expansion of territory. Charles V's rulership was celebrated in Spain and beyond; some even lauded him as the new Roman Emperor. Following Charles V's abdication in 1555–56, Philip II and his ministers aimed to devise "a grand strategy" to extend and justify imperial claims for universal rule. The Battle of Lepanto in 1571, the annexation of Portugal in 1580, and colonial expansion in Asia and Africa buttressed these claims and aspirations.[12]

Spain's imperial expansion and cartography went hand in hand. In fact, the peak of Spanish involvement in maritime and terrestrial cartography coincided with the beginning of Philip's reign. Early in the sixteenth century, the crown established a navigation school attached to the *Casa de la Contratación* (House of Trade) in Seville. Cartographers from Spain, Portugal, and Italy joined with a group of naval pilots and cosmographers whom the crown invited to these schools. These cartographers trained Spanish mariners and mapped each new geographical discovery. Lesser-known schools and institutions also existed in Italy and in other Spanish cities. Philip II further consolidated the Spanish cartographic enterprise by establishing the *Academia de Matemáticas* in Madrid in 1582. Despite the economic and social problems that beset the crown, and the expansion of new centers of cartography in Italy and the Low Countries, Philip continued to support cartography, and Spanish cartographers continued to serve the crown's needs.[13] Eager to keep these maps from his rivals, Philip prohibited the printing of the maps, especially before the 1560s, when he began to institutionalize the production of maps depicting the Iberian Peninsula.[14]

Both the advent of the printing press, and the royal cartography schools in Spain and other imperial centers in Europe, positively affected the production and diffusion of cartographical works. Nevertheless, the medieval tradition of producing *portolan* charts and atlases in small family workshops continued. Charts and atlases, produced in these workshops, continued to focus exclusively on the Mediterranean, even after discoveries had been made in the Indian and Atlantic oceans.[15] Furthermore, practical demand drove the production of *portolan* charts and atlases of the Mediterranean. Since the imperial conflict between the Ottomans and Spanish Habsburgs revolved around the Mediterranean, fine-scale maps of the Mediterranean were useful for both sides, and hence were in demand.

---

[12] Parker, *The Grand Strategy of Philip II*, 11–77.

[13] David Buisseret, "Spanish Peninsular Cartography, 1500–1700," 1072–5.

[14] Ricardo Padrón, *The Spacious World: Cartography, Literature, and Empire in Early Modern Spain* (Chicago: University of Chicago Press, 2004), 40.

[15] In Spain and France, for example, there was a sharp separation between the cartographers who focused on the Atlantic Ocean and those who focused on the Mediterranean. Corradino Astengo, "The Renaissance Chart Tradition in the Mediterranean," in *The History of Cartography, Vol. 3, Bk. 1, Cartography in the European Renaissance*, ed. J. B. Harley and David Woodward (Chicago: University of Chicago Press, 2007), 174–5.

Piri Reis and Ottoman Cartography

In the Ottoman Empire, there was no official cartographer, no cartographical school, no academy devoted to mapmaking. The various maps that Ottoman sultans had at their disposal were the works of individuals, many of whom were former corsairs or naval officers. According to Evliya Çelebi (1611–82), a seventeenth-century Ottoman intellectual and traveler, there was a "Guild of Cartographers" in the capital city in the seventeenth century. Evliya Çelebi maintains that "There are fifteen cartographers working in eight workshops in Constantinople. They are deeply versed in all kinds of sciences and possess different languages, including Latin, in which they read the works of Ptolemy as well as other major works on geography such as *Atlas Minor* and *Mappamonde*. They draw the seas, rivers, mountains of the whole world and sell their maps to sailors and navigators."[16] Another piece of evidence for a palace office for geographers is from Seyyid Lokman's *Şahanşah-name*, prepared for Murad III in 1581. In this illustrated album, one miniature depicts 16 astronomers who are busy drawing maps, working on a globe, and making calculations using astronomical instruments in the first Ottoman observatory. In the first volume of his account, Seyyid Lokman also mentions that intellectuals who worked in the observatory were interested in geography.[17] There are no references to these workshops in other contemporary sources. This does not, however, preclude the possibility that the Ottoman state, like its counterparts in France and Spain, developed an interest in the histories and geographies of different lands, cultures, and peoples.

The Ottoman court was among the first in the world to establish an office of court historian. The Ottomans had been employing historians to formulate their political agenda from the early fifteenth century.[18] The official post of court historian was established around 1550 during the reign of Süleyman the Magnificent.[19] Until very

---

[16] Evliya Çelebi, *Evliya Çelebi Seyahatnamesi*, ed. and trans. Orhan Şaik Gökyay (Istanbul: Yapı Kredi Yayınları, 1995), vol. 1, 236.

[17] Lokman b. Seyyid Hüseyin, *Şahanşah-name*, Istanbul University Library, F 1404, 57a. The relevant part of Seyyid Lokman's work has been edited and translated into English and Turkish by Aydın Sayılı, "Alauddin Mansur'un Istanbul Rasathanesi Hakkındaki Şiirleri," *Belleten* 20 (1956): 411–84. For more information on the observatory and its place within the tradition of observatories in the Muslim world, see Adnan Sayılı, *The Observatory in Islan and Its Place in the General History of the Observatory* (Ankara: TTK, 1960), 289–305.

[18] Fleischer, *Bureaucrat and Intellectual*, 232–52; Douglas Howard, "Ottoman Historiography and the Literature of 'Decline' of the Sixteenth and Seventeenth Centuries," *Journal of Asian History* 22, no. 1 (1988): 52–77; Halil İnalcık, "The Rise of Ottoman Historiography," in *Historical Writing on the Peoples of Asia: Historians of the Middle East*, ed. Bernard Lewis and P. M. Holt (London: Oxford University Press, 1962), 152–67; V. L. Ménage, "The Beginnings of Ottoman Historiography," in *Historical Writing on the Peoples of Asia*, ed. Lewis and Holt, 168–79; Kafadar, *Between Two Worlds*.

[19] Christine Woodhead, "An Experiment in Official Historiography: The Post of Official Şehnameci in the Ottoman Empire, c. 1555–1605," *Wiener Zeitschrift fur die Kundedes Morgenlandes* 75 (1983): 157–82.

recently, the historical literature had argued that the office was a permanent one and that the appointed historian only served the sultan. But Emine Fetvacı's detailed historical analysis on the works and life of Seyyid Lokman, the official historian in tenure from 1569 to 1596/97, has shown that Ottoman court historians did not only write for the sultan but in fact were at the service of a larger circle of court elites. Fetvacı maintains that the office of court historian was not a permanent post, either.[20] Among the main duties of court historians was narrating histories of the Ottoman dynasty and of the universe. But they often commented on contemporary political events as well. In doing so, they articulated the political agendas of their patron and portrayed them in specific ways to fit the imperial vision of the period.

It is safe to assume that the Ottoman court had the same attitude toward geographers and their works. Although there was no official court geographer or cartographer, the sixteenth-century Ottoman court consumed geographical and cartographical accounts at unprecedented levels. Ahmet Karamustafa has argued that in this period, Ottoman cartographical productions fell into two categories: "those drafted under state patronage for administrative use and those produced for private consumption."[21] Included in the first category were architectural plans and siege plans, while works on terrestrial cartography were usually considered private productions.[22] The geographical and cartographical works discussed below belong to both categories. Through the analysis of these works this chapter will also maintain that in the sixteenth-century Ottoman Empire, the border between private consumption and administrative usage was much more fluid than so far believed. The Ottoman court employed geographers and their works in the same way they employed historians. It did not create a permanent post for geographers at the court. However, the elite patrons who sought to shape the political agendas of the court found geographical works and in particular maps not only visually appealing but also very useful instruments to articulate their political visions. Sultans, grand viziers, and princes, who rivaled their siblings during throne struggles, commissioned cartographical and geographical works to consolidate their political agendas and to facilitate the administration of the empire's domains.

In the Ottoman Empire, the person who came closest to becoming an official cartographer was Piri Reis.[23] Piri Reis's contributions to the vision and aims of the sixteenth-century Ottoman Empire were substantial. His works jibed with the Ottoman court and its elite's political and ideological discourse. Piri Reis wrote

---

20 Emine Fetvacı, "The Office of Ottoman Court Historian," in *Studies on Istanbul and Beyond: The Freely Papers*, vol. 1, ed. Robert Ousterhout (Philadelphia: University of Pennylvania Museum of Archeology and Anthropology, 2007), 6–21; Fetvacı, *Picturing History at the Ottoman Court*.

21 Ahmet T. Karamustafa, "Military, Administrative, and Scholarly Maps and Plans," 209.

22 Ibid., 209–27.

23 Svat Soucek, "Piri Reis and the Ottoman Discovery of the Great Discoveries," 133–5.

and gained prestige under the patronage of İbrahim Pasha, who saw control of the Mediterranean as key to attaining universal rule. Moreover, by composing the first comprehensive geographic work in Ottoman Turkish, Piri Reis became a pioneer in promoting Ottoman Turkish as the language of geographic works in the Ottoman Empire. It is fair to assume that Piri Reis inspired Mustafa b. Ali al-Muvakkit, who, in the same period, prepared geographical accounts in Ottoman Turkish for Sultan Süleyman.[24] Finally, Piri Reis's frequent references to his experiences at sea indicate a change in the Ottoman Empire's approach to gathering geographical knowledge, so that traditional geographical accounts were now joined by accounts based on personal experience and the latest knowledge.

The work that best demonstrates Piri Reis's impact on the affairs of the empire is his navigational guidebook, the *Book of Sea Lore*. The *Book of Sea Lore* was presented to Süleyman in 1526 while the sultan was getting ready to challenge Charles V's domains in Europe. The guidebook circulated widely among palace officials from the sixteenth century onward; even today, there are 45 extant copies of the work.[25] Among the surviving copies, some two dozen date from the sixteenth century.[26] With a few exceptions, these copies found their way to the Ottoman palace library, which accounts for the work's popularity, especially within sixteenth-century palace circles.

The *Book of Sea Lore* was an important contribution to sixteenth-century geographical knowledge in general, and to the genre of *isolario* (island book) in particular. Piri Reis's *Book of Sea Lore* was not officially called an *isolario*. It was intended to be and was called a sailor's guidebook. Nevertheless, it served a similar function and had a similar style to contemporaneous *isolarii* produced and circulated in Europe.[27] Historically, *isolarii* had been written for entertainment and consultation on land, rather than for practical use at sea. They portrayed the geographical features of islands in the Mediterranean, and also included classical narratives of the marvelous and monstrous. Today, the Greek historian Diodorus—who described the Mediterranean islands in the fifth book of his *Historical Library*—is considered to be the father of the genre. In the fifteenth

---

24 For Mustafa b. Ali's contribution to the field of geography and Ottoman imperial claims, see the second chapter.

25 The copy, which is located at the Süleymaniye Library in Istanbul under the catalogue number Ayasofya 2612, was donated to the library by Mahmud I (r. 1730–54) and is attested to be the most complete copy. For more information on the extant copies of the *Book of Sea Lore*, see Mine Özen Esiner, *Piri Reis and His Charts*, 20–22.

26 İhsanoğlu, *Osmanlı Coğrafya Literatürü Tarihi*, vol. 1, 23–5.

27 Benedetto Bordone's *Libro ... de tutte l'isole del mondo*, which was published in Venice in 1528, for instance, portrays 111 islands in the Atlantic Ocean, the Mediterranean, and the Far East. *Libro ... de tutte l'isole del mondo* also contains a world map, a map of Europe, and another one of the northeastern Mediterranean. Of the 111 maps, 62 are of the Greek islands and are mostly copied from Buondelmonti and Bartolommeo. Unlike Piri Reis, Bordone doesn't convey his own observations, most likely because he never left northeastern Italy. George Tolias, "Isolarii, Fifteenth to Seventeenth Century," in *The History of Cartography, Vol. 3, Bk. 1, Cartography in the European Renaissance*, ed. J. B. Harley and David Woodward (Chicago: University of Chicago Press, 2007), 270–71.

century, Cristoforo Buondelmonti, a Florentine monk, prepared an *isolario* that became a model for future generations.[28] The extant copies of Buondelmonti's *Liber Insularum Archipleagi* (*Description of the Islands of the Archipelago*) date from 1430 to 1642.[29] The most complete copy includes maps and descriptions of 79 places in the Ionian and Aegean seas, as well as illustrations for Rhodes, Gallipoli, Athens, and Constantinople.[30] Although Buondelmonti drew upon Islamic and Greek sources for his book, pilgrimage guides describing the routes to Byzantium and the Holy Land also influenced his work.[31] The *Liber Insularum* inspired future *isolarii*, which came to include the islands of the Mediterranean and on rare occasions the Atlantic Ocean.

The *isolario* as a genre flourished during the sixteenth century. This was the time when empirical knowledge became an essential part of the learning experience in Europe. In this period, *isolarii* became popular geographical texts that rulers sought to possess as part of their imperial propaganda and out of self-interest.[32] In the early decades of the sixteenth century, two cartographers from opposite ends of the Mediterranean prepared nautical *isolarii* for their respective imperial rulers.[33] Portuguese cartographer Valentin Fernandes prepared his *isolario*, *De insulis et peregrinatione Lusitanorum*, between 1506 and 1510. This work survives in only one manuscript copy, and focuses exclusively on the islands of the Atlantic Ocean. Written during the Portuguese overseas expansion, *De insulis* highlights the new Portuguese territories. Piri Reis's *Book of Sea Lore*, prepared around the same time, focuses exclusively on the Mediterranean and is notable for its great accuracy and detail.[34] Like Fernandes, Piri Reis wrote in the vernacular and thus marked a turning point in both the genre and the development of Ottoman Turkish as a technical language.[35]

---

28 Cosgrove, *Apollo's Eye*, 91.

29 The work appeared in several versions in Rhodes and Constantinople. For the copies in the Ottoman palace library and Mehmed II's interest in the work, see Chapter 2.

30 Tolias, "Isolarii," 265–7.

31 Cosgrove, *Apollo's Eye*, 91.

32 Tolias, "Isolarii," 279–84.

33 There were three main types of *isolarii* in the early modern period: nautical, humanistic, and topical. Piri Reis's and Fernandes's *isolarii* were nautical *isolarii*. Ibid., 281–2.

34 Both Piri Reis and Fernandes were clearly inspired by the Venetian shipmaster Bartolommeo Dalli Sonetti's *isolario* on the Aegean Islands. Bartolommeo's *isolario* was first printed around 1485 and contained 49 island maps. What is important here for our purposes is that the work also contained commentaries about these islands in verse. Bartolommeo's *isolario* did not have a title or a colophon. On Bartolommeo and his *isolario*, see Tolias, "Isolarii," 268–9; Cosgrove, *Apollo's Eye*, 93.

35 Tolias, "Isolarii," 270. Piri Reis's endeavor also coincides with the time when poetry written in Ottoman Turkish began to flourish and become popular in intellectual circles in Istanbul. See Soo Young, "Minding the Shop: Zati and the Making of Ottoman Poetry in the First Half of the Sixteenth Century" (PhD diss., University of Chicago, 2005). Furthermore, Ottoman Turkish gradually developed as the official language in legal affairs and the medical profession.

Piri Reis compiled the initial edition of the *Book of Sea Lore* with simple charts and depictions of the islands in the Mediterranean.[36] In the second edition, he improved the work by adding a total of 214 colorful *portolan* charts that depict various islands and port cities around the Mediterranean.[37] Furthermore, he wrote a new and lengthy introduction—which was composed in rhyming verse—and extended the sections in which he presented anecdotes about local practices, his activities in the region, and naval battles. These detailed descriptions and personal accounts far exceeded those found in contemporary Portuguese, Catalan, and Italian guidebooks.[38] Moreover, these additions made clear to Piri Reis's patron İbrahim Pasha and the Ottoman sultan that Piri Reis was not only a seaman but also a talented intellectual. Herein lay the value of this guidebook for Piri Reis and for the history of geographical knowledge in the Ottoman Empire. Piri Reis was clearly reinforcing his patron's political agenda. But, at the same time, he was also promoting himself as a valuable and prized "geographer" for the Ottoman court.

The *Book of Sea Lore* and the Ottoman Mediterranean

In his introduction, after a long passage on mariners' life on the seas, Piri Reis conveys his knowledge about various topics ranging from the classification of sea storms, to techniques of mapmaking and compass use, to the history of Portuguese discoveries in the Indian Ocean and Columbus's discovery of the New World. Piri Reis discusses Columbus's discoveries in the New World in the text surrounding his 1513 world map.[39] In the *Book of Sea Lore*, Piri Reis further elaborates on this and devotes a short section to the Atlantic Ocean and the Spanish discoveries in this region.[40] The longest section in the introduction is reserved for Portuguese activities in the Indian Ocean and the Red Sea, and their effects on trade; Piri Reis also uses this section to address the difficulties a mariner would face in sailing around the Indian Ocean.[41] With his introduction, which includes information

[36] For the first edition of the *Book of Sea Lore*, see Piri Reis, *Bahrije*, *Das türkisches Segelhandbuch für das Mittelandische Meer von Jahre 1521*, ed. Paul Kahle (Berlin: Walter de Gruyter, 1926). The first version of the *Book of Sea Lore*, which does not include an introduction, opens directly with a description of Tenados, an island located off Gallipoli in the Aegean Sea. Piri Reis, *Bahrije*, vol. 1, 1a.

[37] Piri Reis, *Kitab-ı Bahriye*, Süleymaniye Library, Ayasofya 2612, 4a–43b. Piri Reis, *Kitab-ı Bahriye*, ed. Ökte et al., vol. 1, 46–205.

[38] Soucek, "Islamic Charting in the Mediterranean," 262.

[39] Piri Reis, *Mappamundi* (1513), TKS. Revan Köşkü 1633 mük.

[40] Piri Reis, *Kitab-ı Bahriye*, Süleymaniye Library, Ayasofya 2612, 39a–40b. Piri Reis, *Kitab-ı Bahriye*, ed. Ökte et al., vol. 1, 186–93.

[41] Piri Reis, *Kitab-ı Bahriye*, Süleymaniye Library, Ayasofya 2612, 22a–24b. Piri Reis, *Kitab-ı Bahriye*, ed. Ökte et al., vol. 1, 118–29. On Piri Reis's account of the Indian Ocean and its importance in intellectual and political terms, see Casale, *The Ottoman Age of Exploration*, 37.

on the latest developments in the science of navigation, and on the most recent political events and geographical discoveries in the Indian Ocean and the New World, Piri Reis clearly wanted to impress the Ottoman sultan, who was becoming increasingly open to learning about different parts of the world.

Piri Reis devotes the rest of his work to a subject of great interest and import to the Ottomans: the Mediterranean. In describing this region, Piri Reis suddenly changes his writing style from verse to prose. He explains his reason for this change as follows: "The reason for shifting to prose here is that so far in this book, we have explained the matters regarding compass, map and the situation of the winds and shallows completely in verse. But we described the Mediterranean in prose. Had we done it in verse, it would have taken very long. When you use this book at night or in places of difficulty, verbosity is not proper."[42] By changing his writing style and explaining his reasons for this change, Piri Reis drew the attention of his readers to his main focus: the Mediterranean. The geographer prepared his navigational guide at a crucial time in İbrahim Pasha's imperial enterprise which portrayed the Ottoman sultan as the universal sovereign aiming to control the whole world. Piri Reis, an active sea captain during the Ottoman campaigns that aimed to bring the Aegean Sea under Ottoman control, articulated the political agenda of his patron in his work and helped bring it to fruition by providing important detailed geographical and historical information about the Mediterranean, especially the eastern Mediterranean and the North African coast.

After the introduction, Piri Reis begins the body of the work with a depiction of the Dardanelles, his home region, and from there portrays the rest of the Mediterranean, focusing in particular on the eastern Mediterranean and the North African coast. Of the 214 charts in the second version of the *Book of Sea Lore*, only 32 depict places in the central and western Mediterranean, such as Malta, Sardinia, Corsica, the Balearic Islands of Minorca and Majorca, and Rome. The remaining 182 charts (approximately 85 percent of the total) illustrate the eastern Mediterranean and the North African coast. When it comes to the charts and descriptions of the eastern Mediterranean and the North African coast, the originality of Piri Reis's work is without question. In terms of detail and accuracy, the eastern Mediterranean and North Africa chapters of the *Book of Sea Lore* far exceed anything found in Italian or Catalan *portolans* or *isolarii*. Piri Reis includes more charts for these areas than his contemporaries, and he fleshes out accounts on daily life and on the history of these regions with his own observations.[43]

Surrounding these charts of the eastern Mediterranean and the North African coast are long descriptions of recent history, legends, and instructions for safe travel. Here, Piri Reis describes almost every town, port, and city. Because Piri Reis spent most of his career in these regions, his descriptions here are the most complete.[44] But this is not the only reason. Integrating anecdotes and descriptions

---

42 Piri Reis, *Kitab-ı Bahriye*, Süleymaniye Library, Ayasofya 2612, 43a. Piri Reis, *Kitab-ı Bahriye*, ed. Ökte et al., vol. 1, 202–3.

43 Soucek, "Islamic Charting in the Mediterranean," 272–9.

44 Ibid., 262.

of topography, nature, and the inhabitants of islands into *isolarii* was in fact part of the tradition. In the *Book of Sea Lore*, however, Piri Reis goes beyond the tradition and provides a very detailed account of the economic resources and the social and cultural values of the places explored, which the Ottoman state possibly referred to during the economic and social appropriation of these newly conquered territories. The chapters on Egypt, the Nile, Damascus, and Alexandria, for instance, include minute details about topography—such as coastal vegetation, sources of drinking water, ancient sites, and mountainous areas—as well as useful hints about how to safely navigate these difficult waters. In his chapter on Alexandria, which extends to four folios and is accompanied with a two-folio chart of the city, Piri Reis opens with the history of the city: "It has been told in the histories that Alexandria was built in ancient times. It has been narrated that Alexander the Great found the city in ruins and rebuilt it [...] This city is an important stronghold for the Arab lands."[45] In his chapter on the Nile, Piri Reis explains why he depicts this river and underlines once more the real intention of his work: "In writing about the towns and islands and other landmarks of the Nile, it is not our intention to treat them for practical navigation as we did in the Mediterranean [...] as we already described every place we saw in the Mediterranean, it is necessary to record the places that we saw along the Nile up to Egypt."[46]

Piri Reis's chapters on Djerba, Tunis, and Tripoli are also remarkably rich in details about castles, harbors, the recent history of the region, and local customs.[47] In the following passage, Piri Reis describes the people and the local customs of Djerba:

> The tribe of Sheikh Yahya does not perform its prayers with the *imam* at all, even on Fridays. They don't touch the glass when they drink, and if they do, then they break the glass. There is a Christian cross on the door of every house. When I asked about it, they said that the devil runs away from the cross. They speak a language that the Arabs don't understand. They speak like gypsies.[48]

Similar passages inform the reader of religious practices and institutions in various towns in the eastern Mediterranean. With Ayonaroz (the Kesendire cape) and the island of Patmos, for example, Piri Reis first gives the location of the cape and the island and then describes their monasteries.[49] These passages confirm that Piri

---

45 Piri Reis, *Kitab-ı Bahriye*, Süleymaniye Library, Ayasofya 2612, 350b. Piri Reis, *Kitab-ı Bahriye*, ed. Ökte et al., vol. 4, 1484–5.

46 Piri Reis, *Kitab-ı Bahriye*, Süleymaniye Library, Ayasofya 2612, 355b. Piri Reis, *Kitab-ı Bahriye*, ed. Ökte et al., vol. 4, 1502–3.

47 Piri Reis, *Kitab-ı Bahriye*, Süleymaniye Library, Ayasofya 2612, 333a–336b. Piri Reis, *Kitab-ı Bahriye*, ed. Ökte et al., vol. 4, 1414–29.

48 Piri Reis, *Kitab-ı Bahriye*, Süleymaniye Library, Ayasofya 2612, 333a. Piri Reis, *Kitab-ı Bahriye*, ed. Ökte et al., vol. 4, 1414–15.

49 Piri Reis, *Kitab-ı Bahriye*, Süleymaniye Library, Ayasofya 2612, 56a–56b and 99a. Piri Reis, *Kitab-ı Bahriye*, ed. Ökte et al., vol. 1, 254–7 and 422–3.

Reis was clearly driven by a desire to create an instructive and impressive work on the Mediterranean. He wanted to provide as detailed information as possible about the culture and religion of the inhabitants of the Mediterranean.

In separate chapters, Piri Reis also describes and illustrates important islands and port cities in the central and western Mediterranean, such as Malta, Corsica, Sardinia, Marseilles, and Gibraltar. These sections, however, are usually devoid of any historical or anthropological information. Their chief value rests in the nautical and technical information he provides. For instance, the chapter on Malta opens with the current political situation in the island and then directly delves into the ways to navigate securely around the island: "The island of Malta is under the suzerainty of the Spanish prince [...] it is a low-lying and prosperous place. It is narrated that it has sixty villages. However, it has a well-preserved castle."[50] Another lengthy chapter describes the island of Sardinia in a similar fashion. Here, Piri Reis mentions that he has been to this island with his uncle Kemal Reis. As testimony to his personal experiences in the island, this nine-folio-long chapter offers minute details about how to navigate around the island, which ports are safer, and which landmarks are useful to find one's way.[51] What is different here from his chapters on the ports and towns of the eastern Mediterranean is that even though Piri Reis gives thorough navigational advice, he omits any information about history and the inhabitants of the island, their religion and their cultural practices.

Piri Reis's navigational guide did more than just describe and map different sites in the Mediterranean; it recalled and highlighted the importance of the Mediterranean, especially the eastern Mediterranean and North Africa for Ottoman imperial aspirations. Under the patronage of İbrahim Pasha, Piri Reis on the one hand tried to impress the Ottoman sultan with his knowledge of the area and on the other aimed to influence his political vision. Piri Reis advised the Ottoman court to focus its energies on the Mediterranean, and in particular the eastern Mediterranean and the North African coast. Less than a decade later, his vision was realized. As Chapter 1 has already shown, after the Ottoman-Habsburg conflict in Southeastern Europe ended in stalemate in the 1530s, the Ottoman fleet intensified its presence in the eastern Mediterranean and North Africa under the admiralty of Hayreddin Barbarossa. Piri Reis was the first Ottoman geographer who took the opportunity to introduce these areas to the Ottoman court and helped them formulate their imperial policies.

After the *Book of Sea Lore*, Piri Reis compiled another cartographic work, a world map. Today, only the northwest corner of this map from 1528 survives. After this time, Reis continued to render his services to the Ottoman fleet in the Red Sea, the Indian Ocean, and the Arabian Sea. In 1547, he was appointed admiral of the fleets of Egypt and India, and commanded expeditions against the Portuguese in the

50 Piri Reis, *Kitab-ı Bahriye*, Süleymaniye Library, Ayasofya 2612, 253b. Piri Reis, *Kitab-ı Bahriye*, ed. Ökte et al., vol. 3, 1078–9.

51 Piri Reis, *Kitab-ı Bahriye*, Süleymaniye Library, Ayasofya 2612, 255b–259b and 99a. Piri Reis, *Kitab-ı Bahriye*, ed. Ökte et al., vol. 3, 1086–103.

Red Sea. As a result of a failed campaign in Hormuz in 1552–53, he was executed in Cairo.[52] Nevertheless, the Ottomans remained keenly interested in cartography and geographical works on the Mediterranean in the following decades. Piri Reis articulated a vision that influenced imperial policy during the reign of Süleyman and beyond. His greatest legacy, however, was his contribution to the knowledge of geography in the Ottoman Empire. With him the Ottoman geographers became aware of the importance of integrating the latest knowledge into their accounts rather than simply relying on the traditional accounts. His *magnum opus* in Ottoman Turkish opened the gates for the next generation of Ottoman geographers who tried to impress and influence the Ottoman court.

Mediterranean Cartography: Charting the Core of the World

That Ottoman and Habsburg rulers were both interested in cartography and geography is indicative of three developments in the sixteenth century: a rising awareness among ruling elites of maps and geographical learning; an increasing appreciation of maps as valuable artifacts; and the role of cartography as an essential component of the state building project. As the third volume of the *History of Cartography* demonstrates, mapping served European rulers in the early modern period by enabling them to both record and enhance their political authority.

A striking development in this period was the emerging notion of space. As the boundaries of states expanded overseas, rulers who emphasized territorial sovereignty sought the help of cartographers to visualize and support their "idea of the state as a precisely defined and delimited geopolitical unit."[53] Early in the fifteenth century, circulation of the Latin translation of Ptolemy's *Geographia* in Europe contributed to the understanding that "land could be measured and described in precise, mathematical terms."[54] Both territorial and economic expansion further consolidated the awareness of and demand for space. In this process, maps served to render the space readable and understandable for everyone. This was the beginning of "official cartography."

Starting in the late fifteenth century, Iberian monarchs used sea charts and other geographic devices to help with the planning of their imperial ventures. In the sixteenth century, both states and merchant companies created special institutions and delegated responsibility to specialized officials for the preparation and conservation of cartographic materials. The creation of the office of mapmaker, whether for economic or political reasons, testifies to the fact that cartography was becoming institutionalized by the second half of the sixteenth century. Following the examples of Spain and Portugal, Venice established special offices for mapping and surveying in this period.[55] The Ottomans, however, did not do so.

---

52 On the execution of Piri Reis and the role of "the Indian Ocean faction" in the decision and its implementation see Casale, *The Ottoman Age of Exploration*, 98 and 114.

53 Kagan and Schmidt, "Maps and the Early Modern State," 662–3.

54 Ibid.

55 Ibid., 662–9.

In the absence of an official framework to support the production of maps, individual cartographers—and not specific institutions—stood out in the Ottoman Empire during this period. Piri Reis, for example, became renowned as an expert in navigation and cartography.[56] He was among the group of geographers—a group that also included Matrakçı Nasuh and Seydi Ali Reis—who sought to turn the Ottoman court's attention to the world beyond the Ottomans' domains and immediate frontiers. Piri Reis, Matrakçı Nasuh, and Seydi Ali Reis were pioneer geographers, and each was gifted, ambitious, and experienced. Inspired by their works, the Ottoman court availed itself of additional cartographic works produced and circulated in Europe during this period. The Ottoman interest in such works is evident from the atlases prepared in Ottoman Turkish in different parts of Europe during the second half of the sixteenth century.

On the eve of great naval battles like the Ottoman siege of Malta in 1565, the Ottoman conquest of Cyprus in 1570, and the Battle of Lepanto in 1571, copious amounts of world maps, navigational charts, and *portolan* atlases prepared in Ottoman Turkish circulated in the Ottoman court in Constantinople. Each of these works appears to have been commissioned for the Ottoman sultan or one of his courtiers, or presented to them to curry political or financial favor. One fine example of this is the *Walters Sea Atlas* (ca. 1560) (*WSA* hereafter), which features seven *portolan* charts and a world map reminiscent of those produced for the courts in Europe. Thomas D. Goodrich, who discovered the only copy of this work in Baltimore in 1984, believes that the atlas was probably prepared in Italy for an Ottoman patron. The similarities between *WSA* and contemporary atlases produced in Europe both in style and content led Goodrich and Svat Soucek to the same conclusion.[57] Although we can identify neither the cartographer nor the client, the elaborate illustrations on the charts suggest that the atlas was prepared for artistic purposes and for a wealthy and prestigious customer, possibly a member of the Ottoman court.[58] Like Piri Reis's guidebook, this atlas and its contents point to the growing awareness of a wider world and the role that cartography played in enhancing a deepening appreciation of the world and imperial space at the Ottoman court.

WSA contains seven *portolan* charts, which depict the following (in order): (1) the Black Sea; (2) the Aegean Sea and the eastern Mediterranean Sea; (3) the central Mediterranean and the Adriatic Sea; (4) the western Mediterranean Sea and Iberia; (5) Northwest Europe; (6) Europe, the Mediterranean, and North Africa; and (7) the Indian Ocean, East Africa, and South Asia. The atlas also includes an oval projection world map. The first map (Plate 3) in the atlas encompasses the Black Sea and the Sea of Marmara. While two full and two half wind roses chart

---

56 Soucek, "Piri Reis and Ottoman Discovery of the Great Discoveries," 135.

57 Svat Soucek, "The 'Ali Macar Reis Atlas' and the Deniz Kitabi," *Imago Mundi* 25 (1971): 17–27. Thomas Goodrich, "The Earliest Ottoman Maritime Atlas: The Walters Deniz Atlası," *Archivum Ottomanicum* 11 (1986): 29.

58 Thomas Goodrich, "The Earliest Ottoman Maritime Atlas," 25–50.

the area, five illustrations decorate the map. Among these illustrations, three are rather simple—blue mountains on green hills. The remaining two illustrations are representations of Bursa and Istanbul. Bursa, one of the former Ottoman capitals, is represented as a generic city without any distinctive identifying features. Its image resembles any other imaginary illustration in the atlas. The illustration of Istanbul, however, which is located next to the full wind rose on the right, is the largest and most detailed of all the city panoramas in the atlas. It depicts in detail the historical peninsula, the Galata region, and the coast of Üsküdar. It realistically portrays the panorama of the city, which was dominated by domes and soaring minarets. Clearly, the artist added these minarets and domes to emphasize the city's new identity under Ottoman rule. The Hagia Irene, Hagia Sophia, Fatih Mosque, and the massive Süleymaniye Mosque—distinguished by a golden flag on its minaret—can be easily discerned. The illustration of Istanbul is one of the largest and most detailed depictions in the entire atlas. In addition, there are 33 smaller depictions, each indicating with a red banner the important castles around the Black Sea coast, the Danube River, and the Gulf of Azov. The inclusion of a large and detailed illustration of Constantinople on this map suggests that the producer of the atlas wanted to present the city as the core of an empire that ruled the lands surrounding the Black Sea and the Mediterranean.[59] Although the Ottomans did not face any serious economic and military challenges in the Black Sea, it was still considered an important part of the Ottoman economic network as the backbone of the imperial capital. Therefore it is not surprising to see the Black Sea chart as the first map in an atlas prepared for the prestigious Ottoman patron.

The second map charts the Aegean, the Sea of Marmara, and the eastern Mediterranean as far as the eastern coast of the Gulf of Sidra. In addition to the 32 castle illustrations (including a large one for Damascus) and four blue mountain panoramas, there is a city illustration of Cairo, one of the Ottomans' most prized possessions. The third map of the atlas (Figure 3.1) presents the Italian Peninsula, as well as the islands of Sicily, Malta, Corsica, and Sardinia. In the north, the map covers the area from Nice to Thessalonica, and in the south from Algeria to Libya. There are three forest panoramas, 18 castle depictions, and city views of Tunis, Venice, and Genoa. While the depiction of Tunis is hazy, the cartographer did position the city in the correct location. The illustrations of Venice and Genoa—powerful city-states and commercial rivals of the Ottomans in the Mediterranean—are closer to reality.

The fourth map shows the northern coast of Africa from Casablanca to Tunis, the Iberian Peninsula's Mediterranean and Atlantic coastlines, and the city of Genoa—home to many Spanish financiers—to the east. There are 30 castle illustrations, two forest depictions with blue mountains, and two city illustrations that show Genoa and Algiers—both are quite realistic. The fifth map depicts the British Isles, France, and the Baltic Sea, with illustrations of around 20 castles, five cities, and one green hill. The sixth map (Figure 3.2) is a large-scale compilation

[59] *Walters Sea Atlas*, Walters Art Museum W 660, 1b–2a.

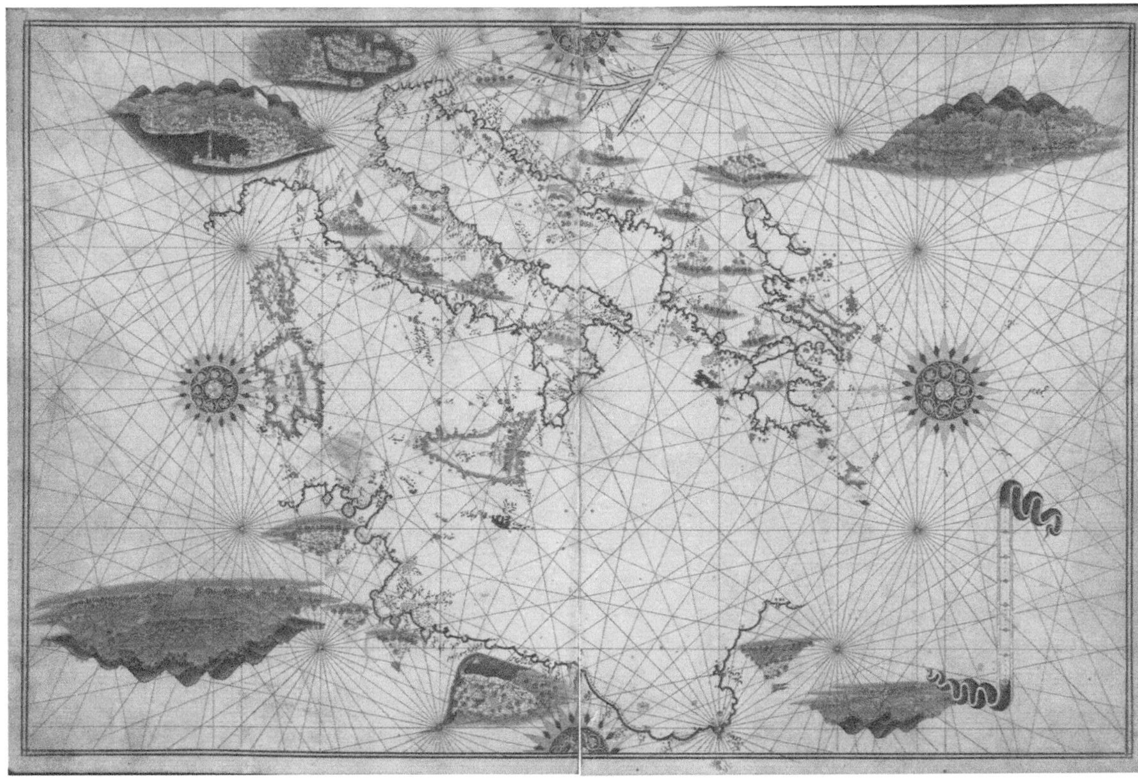

Fig. 3.1 Third map of the *Walters Sea Atlas*. *Walters Sea Atlas.* Walters Art Museum, W. 660, 3b–4a. Courtesy of Walters Art Museum.

(of the first five maps) that links together the various regions—the Mediterranean, the Black Sea, the Sea of Marmara, the Baltic, the Aegean, the Adriatic, the North Sea, parts of the Red Sea and the North Atlantic Ocean, all of Europe, the Canary Islands, and North Africa. There are no illustrations of cities, green hills, or castles on this map. Its aim is simpler than that of the other maps in the atlas—to provide the court with a large-scale map of western Eurasia.

The seventh map (Figure 3.3) charts the Indian Ocean, the eastern third of Africa, and the southern third of Asia on an even larger scale. This single chart has drawn a lot of attention in the historical literature and has been used as significant evidence for rising Ottoman interest in the Indian Ocean.[60] This chart is, however, the only representation of the region in the atlases prepared for Ottoman patrons in the sixteenth century. It is plausible that the targeted audience for this first atlas in Ottoman Turkish was a member of the Ottoman court who was politically and economically interested in the Indian Ocean. The omission of the region from the other two atlases of the same period, however, speaks to the conscious choice by the Ottoman court of ignoring this region of the world. As the empire reached its territorial extent in this period, the ruling elites of the Ottoman court were more interested in sustaining the empire than expanding its territories into faraway and possibly dangerous waters.[61]

Commissioned at the end of Süleyman's reign, *WSA* suggests first of all an ongoing interest at the Ottoman court in maps and atlases, valuable resources that, among other things, enabled the owners to display their prestige and authority. This was a trend that had started during the reign of Mehmed II and was consolidated during the reign of Süleyman with Piri Reis's *Book of Sea Lore*. Second, *WSA* points to a growing awareness among court members of the usefulness of maps. Members of the Ottoman court knew that the Spanish Habsburg court employed maps to elucidate and reflect their imperial aspirations. This understanding gave the Ottoman court an incentive to collect and commission maps of the Mediterranean as well.

The maps and atlases discussed here were just the tip of the proverbial iceberg. Given the nature of Ottoman expansion in the Mediterranean, one can assume that Ottoman seamen routinely used charts, *portolan*s, itineraries, and guidebooks to navigate in friendly and enemy waters, be it for profit, strategic purposes, or conquest. Unfortunately, most of these books and charts must have been lost, which is in part a result of the court's lack of cartographic institutions. As the imperial rivalry between the Habsburgs and the Ottomans focused more and more on the Mediterranean, an Ottoman court elite clearly felt the need to have an atlas that faithfully depicted the Mediterranean.[62] *WSA*—with its six maps that focus

60 Casale, *The Ottoman Age of Exploration*, 192.

61 On the reasons for the Ottoman court's conscious choice of ignoring the rest of the world, in particular the Western Europeans, see Tezcan, "The 'Frank' in the Ottoman Eye of 1583," 267–96.

62 For the importance of *WSA* within the context of Ottoman imperial interest in the Indian Ocean, see Casale, *Ottoman Age of Exploration*, 191–2.

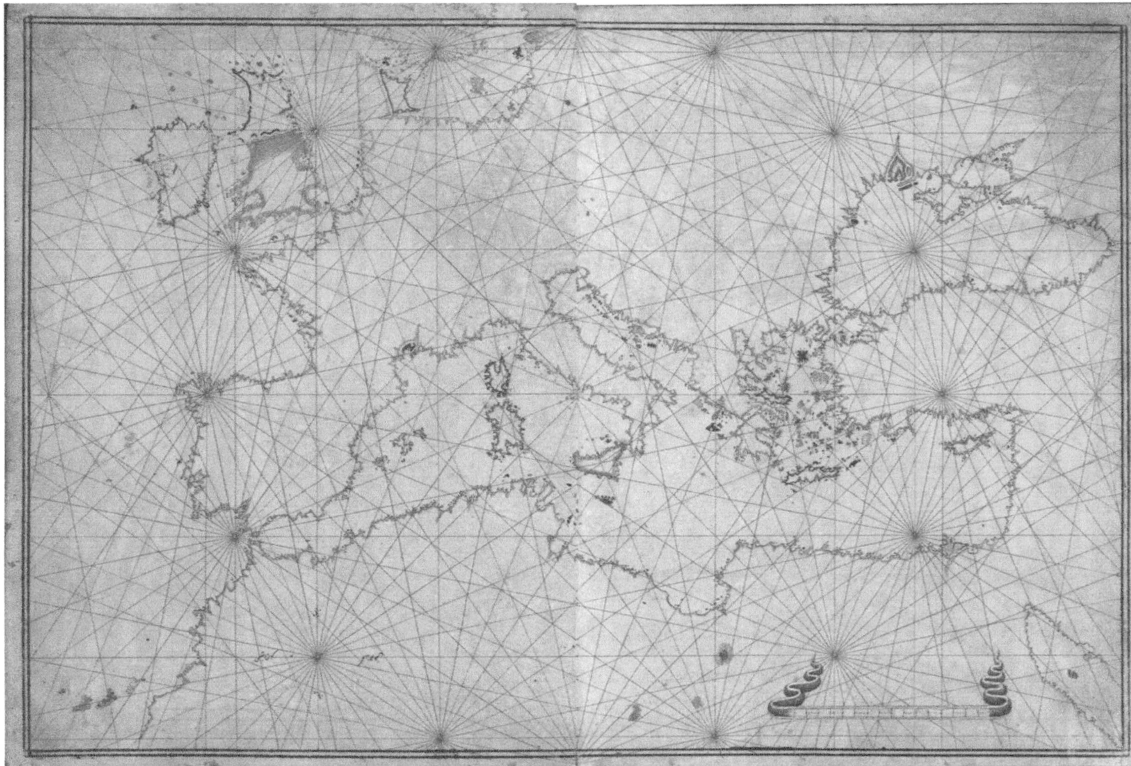

Fig. 3.2 Sixth map of the *Walters Sea Atlas*. *Walters Sea Atlas.* Walters Art Museum, W. 660, 6b–7a. Courtesy of Walters Art Museum.

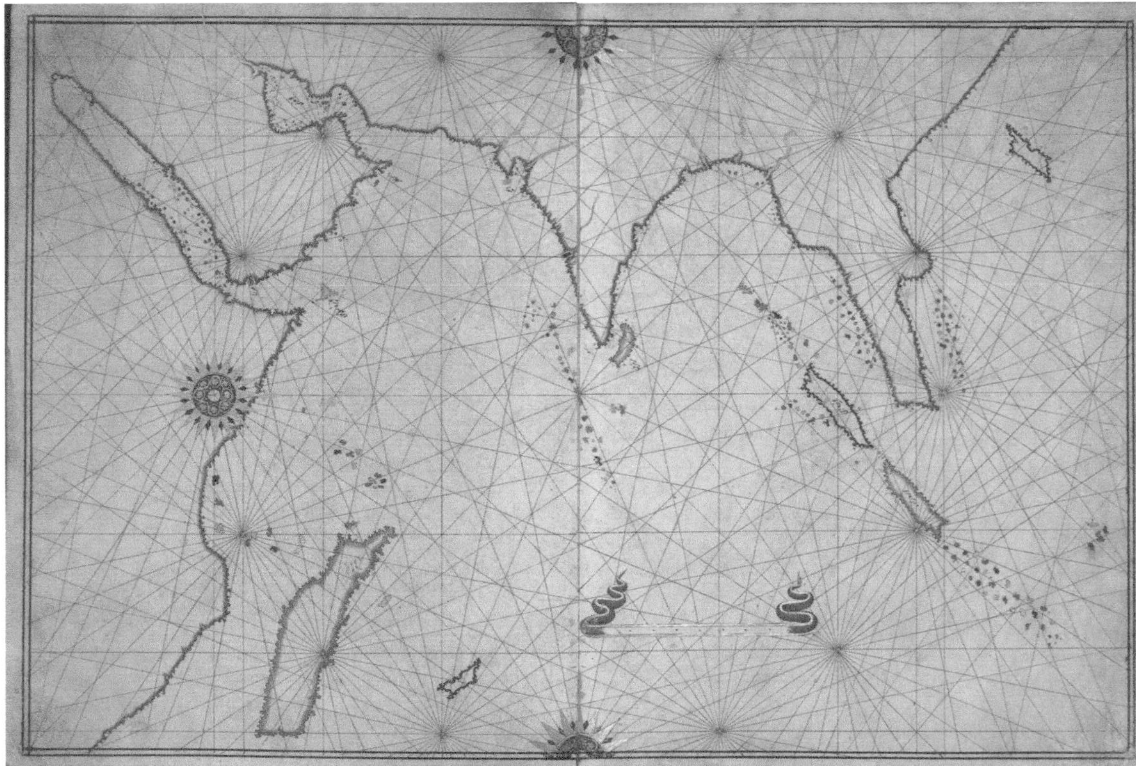

Fig. 3.3 Seventh map of the *Walters Sea Atlas*. *Walters Sea Atlas, Walters Art Museum, W. 660, 7b–8a.* Courtesy of Walters Art Museum.

exclusively on the Mediterranean, the coastline of North Africa, the Black Sea, and the eastern Mediterranean—clearly conveys the value the Ottomans placed on the Mediterranean.

The contents of the atlas and the way in which they are organized to meet the needs of the elite also raise new questions about the place of production for the atlas. If the atlas was prepared in Italy, then how did Italian workshops know about the Ottomans' keen interest in the Mediterranean? The Ottomans did not establish permanent embassies in Europe until the eighteenth century. Before then, the court gathered information about the Venetians and Spanish through its Jewish subjects, through European envoys, and through the staff members of European embassies in Istanbul.[63] During the sixteenth century, one of the major concerns of the Ottoman government was the control of information flowing between Istanbul and various European capitals. European envoys in the Ottoman capital did not speak Ottoman Turkish and were not conversant with Ottoman diplomacy. For the most part, they had to rely on Ottoman interpreters who were usually Greek or Latin subjects of the sultan. Most of these dragomans leaked information to the Venetians or the Habsburgs with the aim of getting the Ottoman court's message out, threatening these states, or misinforming them.[64] In the case of *WSA* and its contents, if the atlas was produced in Italy, the commissioner of the atlas would seem to have successfully conveyed to the Italian workshops the Ottomans' strong interest in the Mediterranean. However, the atlas and its structure also speak to the possibility of local production. The Ottoman court's rising interest in the geography of the Mediterranean attracted individual cartographers to the imperial capital in the sixteenth century. *WSA* might just as well be the product of one these anonymous cartographers.

Two more *portolan* atlases produced in Ottoman Turkish during the same period confirm these points: the atlas of Ali Macar Reis (1567) (*AMR* hereafter) and the *Atlas-ı Hümayun* (*Imperial Atlas*) (ca. 1570). The *AMR* atlas is today preserved at the Topkapı Palace Library.[65] A work of 18 pages, it consists of six nautical charts and a world map prepared in the style of the contemporary Italian schools of Ottomano Freducci and Battista Agnese.[66] The *portolan* charts depict: (1) the Black Sea, (2) the eastern Mediterranean and the Aegean, (3) the Italian Peninsula, (4) the western Mediterranean and the Iberian Peninsula, (5) the British Isles and the Atlantic coast of Europe, and (6) the Aegean Sea, Western Anatolia, and Greece.[67] Ali Macar Reis, an Ottoman sea captain of Hungarian origins, drew

---

63 Ágoston, "Information, Ideology, and Limits of Imperial Policy," 82–3.

64 Ibid. In fact, Ágoston demonstrates that one of these dragomans, Tercüman Mahmud, ordered from Vienna two copies of Abraham Ortelius's *Theatrum Orbis Terrarum* (Theater of the World), the first true modern atlas.

65 Ali Macar Reis, *Atlas* TKS, Hazine 644.

66 Soucek, "Islamic Charting in the Mediterranean," 280.

67 Kemal Özdemir, *Ottoman Nautical Charts: The Atlas of Ali Macar Reis*, trans. P. Mary Işın (Istanbul: Marmara Bank Publication, 1992), 126–30.

the maps of the atlas, which was probably a gift for an Ottoman court elite. It has been suggested that the atlas was actually prepared in Italy and that the place names, which were left blank, were later filled in by Ali Macar Reis.[68] The world map in this atlas is a copy of the large Gastaldi world map of 1561.[69]

AMR provides further insight into the Ottoman court's expanding demand for contemporary maps of the Mediterranean. There are several similarities between *AMR* and *WSA*. The order of the regions in the charts is identical, and the charts in both atlases cover approximately the same areas. Also, the panoramic sketches of the same regions show significant similarities. The first map of *AMR* (Figure 3.4), for instance, like that of *WSA*, is a chart of the Black Sea.[70] Although *AMR* lacks both ornamented wind roses and a detailed illustration of Constantinople, the resemblance between the first maps of each atlas is striking.

The similarities between the two atlases are not limited to the first maps. The second map in *AMR* is also a copy of the second map in *WSA*; it, too, illustrates the eastern Mediterranean and focuses on the Aegean Sea and the North African coast. The third map of *AMR* is again a replica of the third map in *WSA* in terms of the areas it covers. While *AMR* provides a more detailed account of the Adriatic Sea and its islands, *WSA* is focused more on the Italian Peninsula. The fourth map in *AMR* encompasses the same areas as the fourth map in *WSA*. It illustrates the Iberian Peninsula, the peninsula's Atlantic and Mediterranean coasts, the North African coast, and the islands of Corsica and Sardinia. The illustration of a city on the Iberian Peninsula is a replica of the one that appears in *WSA*. The fifth map illustrating the British Isles and the Atlantic coast of Northern Europe is similar to the fifth map in *WSA*. Finally, the sixth map in *AMR* is a large-scale illustration of the Aegean Sea and the Anatolian coast. The only big difference between the two atlases, with respect to the areas they cover, is the omission of the Indian Ocean in *AMR*.

Given that *AMR* replicates *WSA* in large measure, wherein lies its value? Closer study of the regions depicted in the two atlases suggests a possible answer. In both atlases, the Mediterranean (including the Anatolian coastline), the Marmara Sea, the Aegean Sea, and the Adriatic Sea are each depicted seven times. The North African coast and the Atlantic coastline of Europe are each charted twice, and the British Isles and the Black Sea are shown only once. But while *WSA* also includes a chart of the Indian Ocean, *AMR* does not cover this region. The similarities suggest that the Ottomans' understanding of the world was both changing and constant.

---

[68] Soucek, "Ali Macar Reis Atlas," 18–19.

[69] Goodrich, "Some Unpublished Sixteenth-Century Ottoman Maps," in *Comité International d'Études Pré-Ottomanes et Ottomanes, VIth Symposium, Cambridge 1st–4th of July 1984*, eds. J.-L. Bacqué-Grammont and Emeri Van Dunzel (Istanbul: IFEA, 1987), 99.

[70] The illustration in the right corner of a green hill set against mountains, and the depictions of the Danube River (near the Black Sea) and the Gulf of Azov are almost identical to those in *WSA*. See Ian Manners, *European Cartographers and the Ottoman World: Maps from the Collection of O.J. Sopranos* (Chicago: Oriental Institute Museum of the University of Chicago, 2007), 101.

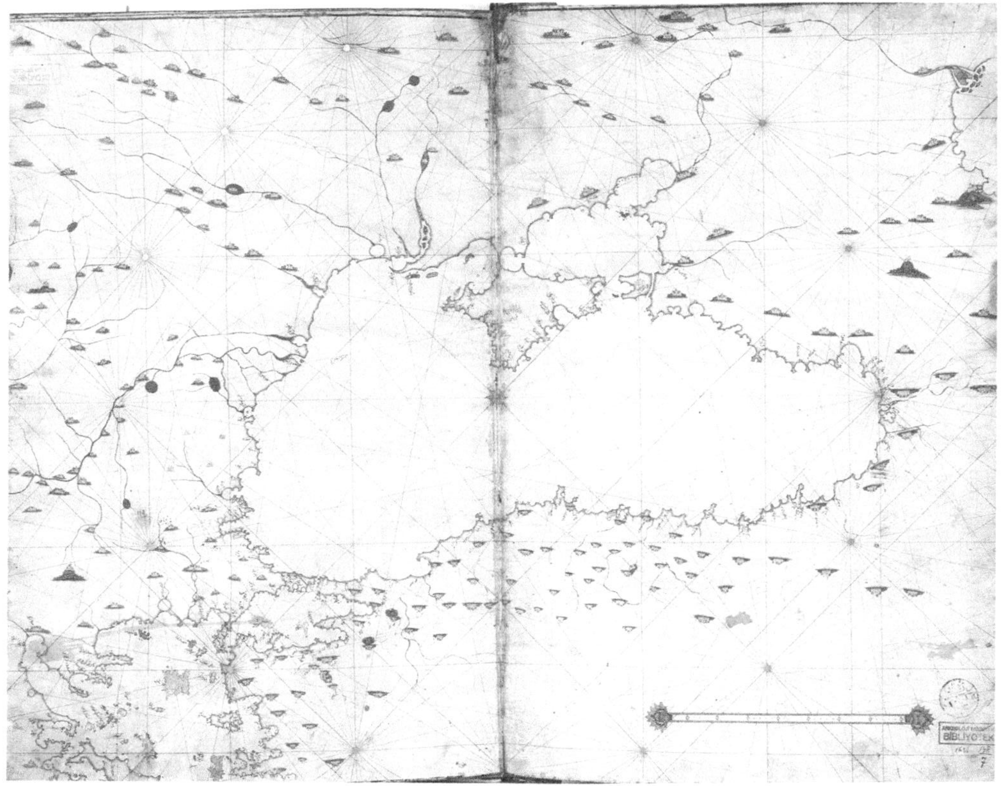

Fig. 3.4 First map of the *Ali Macar Reis Atlas*. Ali Macar Reis, *Atlas*. Hazine 644, 1b–2a. The Library of the Topkapı Palace Museum.

It was constant in that the Mediterranean continued to be of central importance for political, military, and commercial reasons, and evolving in that distant territories, such as the British Isles and the Indian Ocean, could not be ignored.

The *Imperial Atlas*, the last and largest of the atlases prepared in Ottoman Turkish in the sixteenth century, suggests the merits of such a view. Reminiscent of *AMR* and *WSA*, it consists of eight *portolan* charts and a world map. Although there is debate over whether an Ottoman or an Italian workshop produced this atlas, we know for sure that the patron was a member of the Ottoman dynasty.[71] The eight *portolan* charts of the *Imperial Atlas* depict the same areas as the other two atlases: (1) the Black Sea and the Sea of Marmara; (2) the eastern Mediterranean and the Aegean Sea; (3) the central Mediterranean and the Adriatic Sea; (4) the western Mediterranean and Iberia; (5) the central Atlantic coast of Europe and the British Isles; (6) the Aegean Sea and the Sea of Marmara; (7) the Ionian Sea, Greece, and Sicily; and (8) Europe and North Africa.[72]

The *Imperial Atlas* replicates, in general and in detail, *AMR* and *WSA*. But there is one map that makes the *Imperial Atlas* original. The last map of the atlas is a world map that includes Newfoundland and Iceland to the west, the African mainland, the European mainland, the Caspian Sea, and the Arabian Peninsula. The *Imperial Atlas*, in other words, extended the territories far beyond the boundaries of the empire to include much of the world known to the Europeans in 1570. Yet some elements remain constant. The eight maps of the *Imperial Atlas* cover the Mediterranean eight times, while North Africa and the Sea of Marmara are each depicted four times. Moreover, the Black Sea and the Atlantic coast of Europe are charted twice, and the European mainland and the British Isles only once. The *Imperial Atlas* clearly centers on the regions around the eastern Mediterranean Sea. While the Indian Ocean is omitted entirely, the atlas pictures the frontiers of the Ottoman domains, such as the Arabian Peninsula and the Caspian Sea, none of which the other atlases depict. This atlas is particularly interesting not only because of the areas it includes but also because of the outdated information it transmits about the Arabian Peninsula, the Red Sea, and the South African coast. While the cartographer charts the Mediterranean coastal regions in great detail, the other areas mentioned above are represented with distorted images, which are based on at least one century-old cartographic source. This carelessness in depicting the regions beyond the Mediterranean again suggests the relative lack of interest in the frontiers of the world, and the importance of the Mediterranean for members of the Ottoman court, who were the patrons of this work.

Analyzed together, these atlases from the second half of the sixteenth century speak to the Ottoman court's rising awareness of the expanding world. Furthermore, they emphasize the centrality of the Mediterranean in the Ottoman world order. The structure of these atlases reflects the Ottomans' imperial aspirations, even

---

[71] For differing opinions on the place of production of the *Ali Macar Reis Atlas* and *Imperial Atlas*, see Goodrich, "Atlas-ı Hümayun: A Sixteenth-Century Ottoman Maritime Atlas Discovered in 1984," *Archivum Ottomanicum* 10 (1985): 83–101; Soucek, "Ali Macar Reis Atlas."

[72] Goodrich, "Atlas-ı Hümayun," 89–91.

at a time when Ottoman territorial expansion had begun to slow. The first chart, common to all three atlases, is a map of the Black Sea. Despite its proximity to the capital city, the Black Sea appears only one time in each atlas. Atlases prepared for circulation in Europe during the same period commonly place this region at the end of the work.[73] The Black Sea was the "hinterland" of Constantinople, and it was a familiar region for the Ottomans. After Mehmed II's conquests along the Black Sea coast to the north and the south, there was constant commercial traffic between the major ports of the northern Black Sea and the Ottoman capital. This alone might explain why these atlases open with a chart of the Black Sea.[74] But why—if the sea held such importance for the Ottomans—did it show up less often than any other region? The Black Sea was rarely represented in the sixteenth-century *portolan* atlases prepared for consumption in Europe as well. The Black Sea was a safe region for the Ottoman Empire in the sixteenth century. No hostile navies sailed its waters in this period to challenge the Ottomans' power and authority in the region. The Ottomans, like the Bourbons and the Habsburgs, viewed the Mediterranean as the center of the world and the Ottoman court simply considered the Black Sea, which was a safe haven for the Ottoman fleet, a sub-region within the Mediterranean. This may well explain why, although the Black Sea always appeared first, it appeared least often. Since all three atlases were either prepared for or commissioned by the Ottoman court, it is not surprising that they begin with a map depicting the heart of the Ottoman world, the imperial capital city of Constantinople.

The remaining charts depict the Mediterranean more than any other part of the world. The reasons are clear. The Mediterranean was of vital importance to the Ottomans and hence control of its shores was crucial. Without controlling the Mediterranean, the Ottomans could not legitimately claim to be the heirs of Rome, and thus the rightful rulers of a world empire. Less attention is paid to the more distant regions lying on the periphery of the empire, including the British Isles, the Atlantic Ocean, and the Indian Ocean. The omission of the Indian Ocean and any region beyond the Mediterranean did not contradict the Ottoman claims for universality. By denying these regions geographical importance, the Ottoman court actually projected these regions to the peripheries of the Ottoman world and denied the European accomplishments in these regions any importance in the rivalry for universal claims.[75] On the other hand, the inclusion of such outlying regions, even peripherally, testifies to the Ottomans' awareness of the expanding world. That said, these regions are acknowledged almost in passing. Clearly, the interest of the patrons of these atlases lay in the Mediterranean. After admiring the *WSA*, which was the earliest of the three atlases, two other courtiers commissioned similar atlases. These three court elites wanted to display their affluence and to influence

[73] For instance, an atlas prepared in 1554 by Battista Agnese for the abbot of the Benedictine Monastery in Arras features the Black Sea as the thirteenth and final map.

[74] Halil İnalcık, "The Ottoman State: Economy and Society, 1300–1600," in *An Economic and Social History of the Ottoman Empire*, ed. Halil İnalcık and Donald Quataert (Cambridge: Cambridge University Press, 1994), 271–315.

[75] Tezcan, "The 'Frank' in the Ottoman Eye of 1583," 279–80.

imperial decision-making during a period of great battles in the Mediterranean. As Chapter 1 has shown, the Ottoman military engagements in the Indian Ocean were already fragile in the 1550s and completely died down by the time these atlases entered the Ottoman palace. These three atlases were perhaps utilized to draw the attention of the Ottoman sultan and his court back to the Mediterranean.

It has recently been suggested that these three atlases were original contributions to the field of cartography since they were new forms of cartographic representation, transmitted simultaneously from the East and the West in the Mediterranean's fluid structure of intellectual life.[76] It is, however, important to remember that there was a strong *portolan* chart tradition among the Muslim cartographers preceding the Ottomans. The *portolan* charts of the Mediterranean and the Black Sea prepared by al-Katibi from Tunis in 1412–13 and by İbrahim al-Mursi in 1461 reached the Ottoman palace during the reign of Bayezid II.[77] The Topkapı Palace Library in Istanbul today also houses a *portolan* map created by the Majorcan master Johannes de Villadestes in 1428.[78] Ottoman seamen, as Piri Reis indicated on his world map, were also aware of the Portuguese and Spanish charts circulating in the sixteenth-century Mediterranean.

After the 1580s, it became clear that neither the Spanish nor the Ottomans would be able to establish supremacy in the Mediterranean. In 1587, Murad III refused to renew the truce with Spain; however, the Ottoman court was still reluctant to engage in conflicts in the Mediterranean. During this period, the court still welcomed cartographical works on the region. The *Aegean Sea Chart* (1590–91) made by Mehmed Reis of Menemen is an example of a work produced at this time. Debate surrounds both Mehmed Reis's biography and his contribution to Ottoman cartographic tradition. It is equally unclear who commissioned the *portolan* map, which is today located in Venice. It has been suggested that Mehmed Reis was an Ottoman ship captain who acquired a blank chart from a workshop in Istanbul, made some corrections, and then filled in the place names.[79] The chart provides

---

[76] See Tarek Kahlaoui, "The Depiction of the Mediterranean in Islamic Cartography (11th–16th Centuries): The *Suras* (Images) of the Mediterranean from the Bureaucrats to the Sea Captains" (PhD diss., University of Pennsylvania, 2008), 292–321. On the fluid nature of the Mediterranean in the early modern period, especially with respect to the exchange of maps and geographical productions, see Brotton, *Trading Territories*, 87–98. For a criticism of this approach, see Sean Roberts, *Printing a Mediterranean World: Florence, Constantinople, and the Renaissance Geography* (Cambridge, MA: Harvard University Press, 2013), 133–71.

[77] Soucek, "Islamic Charting in the Mediterranean," 264–6; İhsanoğlu, *Osmanlı Coğrafya Litertatürü Tarihi*, vol. 1, 3–4.

[78] Casale, *The Ottoman Age of Exploration*, 18.

[79] On Mehmed Reis's chart and the different opinions on its original cartographer, see Mirco Vedovato, "The Nautical Chart of Mohammed Raus, 1590," *Imago Mundi* 8 (1951): 49; William Brice, Colin Imber, and Richard Lorch, *The Aegean Sea-Chart of Mehmed Reis Ibn Menemenli, A.D. 1590/1* (Manchester: University of Manchester Press, 1977); William Brice and Colin Imber, "Turkish Charts in the 'Portolan' Style," *Geographical Journal* 144 (1978): 528–9; Soucek, "Islamic Charting in the Mediterranean," 283.

great detail about the eastern Mediterranean, covering the Aegean archipelago and circumnavigating the Peloponnese from Durazzo (Albania) to the Dardanelles, the Sea of Marmara, and finally Constantinople. It also charts the Mediterranean coast of Asia Minor up to the Gulf of Alexandretta. Although it contributes very little to cartographic knowledge, the map of Mehmed Reis deserves our attention. If he did indeed buy a blank chart and make certain corrections to it, this suggests that there was still a market for such charts in the Ottoman Empire at this time. Similarly, one fine example of the map market in Constantinople is Al-Hacc Abu'l- Hasan's *portolan* chart of the Mediterranean. Drawn by a seaman from North Africa, this elegant and colorful rendition of the Mediterranean on deer skin possibly reached the Ottoman palace library at the end of the sixteenth century.[80]

While, with the exception of Piri Reis's guidebook, it is hard to determine how many of the cartographical works discussed in this chapter reached a large circle of readers in Istanbul, the court elite—often themselves patrons—were very familiar with these sorts of maps. The availability and the demand for such cartographical works in Istanbul bespeak a constant search by the Ottoman court and individual mapmakers for the latest geographical knowledge about the Mediterranean. The Mediterranean stood in the middle of the Ottoman realm, and it was the stage of many battles. Maps of the region were used both to aid the Ottomans in battle and to reaffirm the Mediterranean's central role in their pursuit of world domination. The region played a significant part in the formation of sixteenth-century Ottoman imperial ideology, which portrayed the Ottoman Empire as a world empire and Ottoman sultans as universal sovereigns. This vision, which was first formulated by Piri Reis, outlived Sultan Selim and Sultan Süleyman, and left an ongoing legacy for future Ottoman rulers.

---

[80] Al-Hacc Abu'l- Hasan, *Portolan Chart*, TKS. Hazine 1822. The confusion about the date of this chart still prevails today. Although İhsanoğlu dates the chart to the seventeenth century, it is cataloged as an example of sixteenth-century *portolan* charts by Fuat Sezgin. İhsanoğlu, *Osmanlı Coğrafya Literatürü Tarihi*, vol. 1, 114; Fuat Sezgin, *Geschichte des Arabischen Schrifttums, Band XII, Mathematische Geographie und Kartographie im Islam und ihr Fortleben im Abendland, Kartenband* (Frankfurt am Main: Institut für Geschichte der Arabisch-Islamischen Wissenschaften an der Johann Wolfgang Goethe-Universität, 2000), 77.

Chapter 4
Projecting the Frontiers of the Known World

> This is a vast sea [Atlantic Ocean], previously, no one knew what was on it. Now they have discovered places on its far shores, for that is how God created it [...] They were never able to discover its limits, though they supposed that they must exist. So they agreed that it must be a single place thus the dispute ended [...] According to the historians, it was in the year 870 of the Hijri calendar when these lands were discovered. They called them the Antilles.[1]

This passage by Piri Reis epitomizes the Ottoman stance towards the distant regions of the world in the sixteenth century. Piri Reis's account informs its readers about the recent discoveries in the Atlantic Ocean, but it does so in a concise yet careless manner. For instance, it summarizes the intellectual disputes about the landmasses bordering the Atlantic Ocean, but it dates the early geographical discoveries in the Atlantic Ocean to 1465–66, at least 26 years earlier than they actually took place. Sixteenth-century Ottoman political and intellectual circles observed the recent geographical discoveries in the Atlantic Ocean closely, however, they consciously located these regions on the borders of their imperial vision and geographical knowledge. This chapter maintains that Piri Reis's account was an early example of a larger corpus of geographical works reflecting a new era in Ottoman history. In the second half of the sixteenth century, Ottoman curiosity and intellectual attention to distant regions of the world increased considerably. In a period when the ruling elite's efforts focused more on the consolidation of Ottoman sovereignty in the regions already under the empire's control, some members of the Ottoman court commissioned geographical works on the far regions of the world in order to change and shape the imperial policies of the Ottoman state.

In the 1550s, Seydi Ali Reis, admiral of the Red Sea fleet, prepared two accounts of geographical and historical value for the Ottoman court: one on India and the other on the Indian Ocean. The first was *Kitab-ı Muhit* (*Book of the Ocean*), which the author completed in 1554 in Ahmedabad—a major town in the Gujarat region of India—and dedicated to Sultan Süleyman.[2] This navigational

---

[1] Piri Reis, *Kitab-ı Bahriye*, 39b. Piri Reis, *Kitab-ı Bahriye*, ed. Ökte et al., vol. 1, 189. Piri Reis here states the date of the discovery of the Antilles wrong. The year 870 of the Hijri calendar corresponds to 1465–66.

[2] There are four manuscript copies of this work, two of which are located in Istanbul. One of them is in the Topkapı Palace Library. Seydi Ali Reis, *Kitab-ı Muhit*, TKS, Revan Köşkü 1643. This copy is believed to be the autograph copy of the work. It was published by Fuat Sezgin in a facsimile edition in 1997. Seydi Ali Reis, *Book of the Indian Ocean: Kitab al-Muhit*, ed. Fuat Sezgin (Frankfurt: IGAIW, 1997). Another manuscript copy of the work from 1666 is in the Nuruosmaniye Library. Seydi Ali Reis, *Kitab-ı Muhit*, Nuruosmaniye Library, 2948. There are two more manuscript copies of the work, one in Vienna and one

guidebook on the Indian Ocean also described in detail the climate and topography of the region. In 1557, Seydi Ali Reis composed the other work, an account of his travels from the port of Surat to Istanbul entitled *Mir'atü'l-Memalik* (*Mirror of Lands*). Prepared for the grand vizier, Rüstem Pasha, *Mirror of Lands* is the most comprehensive account on India written in Ottoman Turkish discovered so far. In this account, the author relates his observations on Indian, Persian, and Crimean societies and rulers, as well as on the geography of these areas.[3]

Working on both Ottoman and Portuguese documents, Giancarlo Casale has argued that Seydi Ali Reis's works were important examples of "synthetic" works by the Ottoman geographers who "repackaged" the learning of Arabic geographers and presented it "in a manner that was both more relevant and more easily digestible for a contemporary Ottoman audience."[4] According to the author, Seydi Ali Reis and his works epitomize the fact that the "Ottoman horizons had expanded far beyond the shores of the Mediterranean" into the Indian Ocean.[5]

This chapter argues that Seydi Ali Reis and a number of Ottoman geographers introduced these different regions and peoples into Ottoman geographical knowledge using both traditional and novel means. They recast the available knowledge about the geography and peoples of India, the Indian Ocean, and the New World. In most cases, even though the geographers benefited from traditional styles or knowledge, they always endeavored to include the latest knowledge available and incorporate their own experiences where possible. In so doing, they created a body of literature that articulated the political agendas of their patrons and in return hoped to shape the imperial policies of the Ottoman court. This chapter analyzes the Ottoman geographical works on the Indian Ocean and the New World as part of a dialogue with contemporary European accounts. It investigates why members of the Ottoman court showed an increasing interest in the Indian Ocean and the New World at this time, how Ottoman geographers responded to this demand, and how the geographers and the court established links to these regions. Careful examination of the production and circulation of these geographical works will reveal that the leading political vision of the sixteenth-century Ottoman Empire projected these regions of the world to the peripheries of the Ottoman world.

---

in Naples. The Vienna copy was written in Diyarbakır in 1558 from the original copy left there by Seydi Ali Reis. Seydi Ali Reis, *Kitab-ı Muhit*, Vienna, Kat. Flügel, nr. 1277; Napoli, Biblioteca Nazionale. On the manuscript copies of *Book of the Ocean*, see İhsanoğlu, *Osmanlı Coğrafya Literatürü Tarihi*, vol. 1, 37–8. On Seydi Ali Reis and his works, see Soucek, "Sidi Ali Reis," *EI*^{2}; C. Orhonlu, "Seydi Ali Reis," *Tarih Enstitüsü Dergisi* 1 (1970): 39–56. On Seydi Ali Reis's career in the Indian Ocean, and on Rüstem Pasha's patronage, see Casale, *The Ottoman Age of Exploration*, 85–6, 88, 120–23, and 186–7.

3 Seydi Ali Reis, *Mir'atü'l-Memalik*, TKS. Revan Köşkü 1470; TKS Koğuşlar 893; TKS. Revan Köşkü 1647; Paris Bibliotheque Nationale, 148; R. Biblioteca Torino DC, Ex Bibliotheca Regis Karoli Alberti, 56. There is also a printed version of the work from the nineteenth century: Seydi Ali Reis, *Mir'atü'l-Memalik* (Istanbul: İkdam Matbaası, 1313/1895).

4 Casale, *The Ottoman Age of Exploration*, 186–7.

5 Ibid., 88, 120–23, 186–7.

India and the Indian Ocean: Ottoman Peripheries to the East

In the fifteenth century, the Portuguese circumnavigated Africa and reached the Indian Ocean. These expeditions, which marked the beginning of Portuguese territorial and mercantile expansion in Asia and the Indian Ocean, did not challenge the early modern geographical understanding of the world. In fact, the Portuguese were latecomers to the Indian Ocean. Merchants from the Italian Peninsula had frequented Indian ports since ancient times, recording the details of their voyages from the thirteenth century onwards. Following the rapid expansion of Islam in the eighth and ninth centuries, Arab sailors, merchants, and travelers frequently ventured into the region and challenged the Ptolemaic understanding of the Indian Ocean as a landlocked sea. Nevertheless, when Vasco da Gama entered the Indian Ocean in 1498, he opened up new horizons for future European sailors. The impact of this discovery was as important as Columbus's 1493 voyage to the Americas. By unlocking the southeast trade networks, the Portuguese opened some of the world's most lucrative trade routes for the rest of the "Age of Exploration," and ushered in Portuguese overseas expansion.[6] From a European perspective, this development had merit. Yet, trade between the Indian Ocean and Far East and the Mediterranean was a centuries-old phenomenon. What changed after 1498 was the shipping route.

The Portuguese territories in the Indian Ocean and in India soon became a source of controversy between Portugal and Spain.[7] In 1481, the papal bull *Aeterni Regis* designated the regions south of the Canaries and west of Africa as exclusively Portuguese. In 1493, another papal bull granted Spain sovereignty over the territories discovered (or to be discovered) by Columbus. The status of Africa and India remained unclear. In 1494, with the Treaty of Tordesillas, Africa and Asia were allocated to the Portuguese, and the western oceanic island world was allocated to Spain. By the 1520s, new geographical discoveries required the extension of the Tordesillas line around the globe. In 1529, Spain and Portugal negotiated a treaty in Saragossa that allowed the Spice Islands to be shifted summarily to either side of the line in future negotiations.[8] The cartographers responsible for collating knowledge on new geographies played crucial roles in the settlement of these controversies.[9]

---

6 Felipe Fernández-Armesto, "Portuguese Expansion in a Global Context," in *Portuguese Oceanic Expansion, 1400–1800*, ed. Francisco Bethencourt and Diogo Ramada Curto (Cambridge: Cambridge University Press, 2007), 495; Kevin Joseph Sheehan, "Iberian Asia: The Strategies of Spanish and Portuguese Empire Building, 1540–1700" (PhD diss., University of California, Berkeley, 2008), 24–72.

7 Sheehan, "Iberian Asia."

8 Cosgrove, *Apollo's Eye*, 84–5.

9 Maria Fernanda Alegria, Suzanne Daveau, João Carlos *Garcia*, and Francesc *Relaño*, "Portuguese Cartography in the Renaissance," in *The History of Cartography, Vol. 3, Bk. 1, Cartography in the European Renaissance*, ed. J. B. Harley and David Woodward (Chicago: University of Chicago Press, 2007), 996–7.

The Portuguese crown was aware of the practical and symbolic value of cartographic accounts, especially world maps, in articulating its imperial claims. As early as the fifteenth century, it established institutions to regulate overseas trade and administration in India, and employed cartographers to oversee map production and nautical science.[10] The *Armazém da Guiné e Índia*, which already existed at the end of the fifteenth century, systematically updated nautical charts so that they reflected the latest discoveries.[11] Portuguese rulers also continued to commission cartographers to draw world maps for them. In 1457, Dom Henrique's (Henry the Navigator) nephew, Afonso V, commissioned the Italian cartographer Fra Mauro to draw a world map, which incorporated the details of Portuguese voyages along the west coast of Africa.[12] Portuguese geographic accounts, in particular cartographical production on the Indian Ocean and India, diminished in the second half of the sixteenth century. After 1580, increasing Dutch interest in the Indian Ocean, and the united Spanish and Portuguese crown's need to maintain and control the coastal zone of India, spurred the production of military and urban cartography.[13] These maps incorporated knowledge from both eyewitness accounts and accounts written before the European discoveries.

Ottoman political and commercial interest in the Indian Ocean and India was not as systematic or centralized as in the case of the Portuguese. As Chapter 1 has shown, as soon as they reached the shores of the Indian Ocean in the early sixteenth century, the Ottoman state established a clear yet cautious presence in the region. Ottoman ruling elites were well aware of the economic importance of the Indian Ocean, which played a significant role in the spice trade. Ottoman efforts to take control of the Red Sea and its trade network had already started during the reign of Selim I. Selim's eastern campaigns and, later, the conquest of Egypt and the Hejaz region brought the Ottomans to the shores of the Red Sea. With conquest came a sense of responsibility to protect Ottoman ships and domains in this area.

During the reign of Sultan Süleyman, the Red Sea and the Indian Ocean became arenas for different factions of the Ottoman court to rival each other and display their political and military talents and skills.[14] The geographical distance of the Indian Ocean worked against its political and economic significance in the eyes of many at the Ottoman court, including Süleyman and his grand vizier İbrahim Pasha, whose main concern in this period was clearly the Mediterranean. Nevertheless, the Ottomans aimed to thwart Portuguese efforts to expand in the

---

[10] For an account of the efforts of the Portuguese crown to collect and standardize cartographic information, see David Turnbull, "Cartography and Science in Early Modern Europe: Mapping the Construction of Knowledge Spaces," *Imago Mundi* 48 (1996): 5–24.

[11] Alegria, Daveau, Garcia, and Relaño, "Portuguese Cartography in the Renaissance," 1003–4.

[12] Brotton, *Trading Territories*, 50–51.

[13] Alegria, Daveau, Garcia, and Relaño, "Portuguese Cartography in the Renaissance," 1019–22.

[14] Casale, *The Ottoman Age of Exploration*.

region surrounding the Red Sea, which required solid control of the southern coast of the Arabian Peninsula. At the same time, they endeavored to establish political and economic alliances with the Muslim communities in Gujarat. The Ottoman fleet, especially under the command of Hadım Süleyman Pasha, sometimes engaged in the region to challenge the Portuguese militarily. The Ottoman economic ambitions in the region also targeted Portuguese merchant ships, but ideologically the Ottoman court also wanted to present a formidable challenge to the Safavid shahs' claims to be the rulers of Muslims the world over and keep the Hejaz region safe. This political agenda was strongly challenged by the Ottoman geographers who tried to transmit the latest knowledge about these regions to the Ottoman court and who, in doing so, aimed at formulating and reformulating the court's imperial ideology in the Indian Ocean.

India and the Indian Ocean in Sixteenth-Century Ottoman Geographical Knowledge

Ottoman intellectuals, particularly geographers, had been fascinated with the Indian Ocean's geography since the early decades of the sixteenth century. The earliest extant geographical account that informs the Ottoman court of political developments in the Indian Ocean is the introduction of Piri Reis's *Book of Sea Lore*. In the introduction of his work's second version from 1526, Piri Reis discusses the voyages of Portuguese seamen in the Indian Ocean, and how the Portuguese came to establish control in the region. In this section, Piri Reis displays his familiarity with navigational techniques and instruments used by the Portuguese, such as astrolabes, hourglasses, maps, and the knowledge of calculating parallels. Yet, the recurrent theme of his narrative is the dangers of the Indian Ocean. Piri Reis cautions his readers about the difficulties in navigating this "infinite" (*bi-giran*) ocean and almost goes so far as to advise Ottoman seamen to stay out of the region, if possible. Moreover, Piri Reis does not hide his admiration for the Portuguese, who managed to sail through the strong storms and heavy rains that rendered navigating the Indian Ocean dangerous.[15] Perhaps it is not a coincidence that Piri Reis's only failure as a sea captain took place in the Indian Ocean. The prestigious captain, admiral of the Indian fleet, and geographer, whom the Ottoman court admired and respected on account of his geographical works and service in the Mediterranean, failed to capture Hormuz in 1552. Feeling very unconfident in these not-so-friendly waters, the admiral abandoned his fleet in Basra and fled to Suez. Piri Reis paid for this disgraceful deed with his life; he was executed by the Ottoman officials.

The Grand Vizier Rüstem Pasha charged Seydi Ali Reis with taking the abandoned armada from Basra back to Suez. Seydi Ali Reis came from a well-established family of Black Sea seamen. He served in the Ottoman navy during

[15] Piri Reis, *Kitab-ı Bahriye*, 22a–24b. Piri Reis, *Kitab-ı Bahriye*, ed. Ökte et al., vol. 1, 118–29.

the conquest of Rhodes and joined Hayreddin Barbarossa in other Mediterranean campaigns. Under the patronage of the Grand Vizier Rüstem Pasha, the sea captain also became an active participant in Constantinople's intellectual and political life. Despite having spent most of his career in the Mediterranean, Seydi Ali Reis's repute as a successful seaman and a prominent geographer stems from his service in the Indian Ocean.[16] In 1553, Sultan Süleyman appointed him admiral of the Red Sea fleet. His first mission in this post was to take to Egypt the Ottoman naval ships that Piri Reis had abandoned in Basra. After being ambushed by the Portuguese off the coast of Muscat, the majority of Seydi Ali's ships either sank or were captured by the Portuguese. Seydi Ali Reis was unable to take the remaining ships back to Suez and instead anchored them in the port of Surat in Gujarat.

Seydi Ali Reis's mission in the Indian Ocean yielded two important geographic works on the Indian Ocean and India. It was while he was in Gujarat in 1554 that Reis wrote the *Book of the Ocean*, a sailor's guide for the Indian Ocean. For decades, scholars treated the *Book of the Ocean* either as a simple translation of several Arabic navigation manuals or as a complementary work to Piri Reis's *Book of Sea Lore*.[17] Only recently has Giancarlo Casale reevaluated the work and presented it as important evidence of the Ottomans' imperial interest in the Indian Ocean.[18]

Although his experiences clearly informed his account, it seems that Seydi Ali Reis also benefited from accounts on the Indian Ocean by Ibn Majid (d. ca. 1500), the most famous and one of the last Indian Ocean navigators prior to the arrival of the Portuguese, and Sulaiman al-Mahri (d. ca. 1554), another famous Arab navigator.[19] The work opens with a dedication to Süleyman I in which Seydi Ali Reis refers to the Ottoman sultan as the "ruler of the seven climes" and the *afitab-ı 'alem-tab* (sun illuminating the universe).[20] Seydi Ali Reis then promptly explains the title of his book and expresses his hopes for the reception of this navigation manual: "This book is called *Muhit*[21] because it covers all the oddities of the science of navigation in the seas. I hope that those knowledgeable people who read it can correct the mistakes with the pen of forgiveness. I also hope that those who benefit from the book in sailing remember its author and pray for him."[22]

---

16 Soucek, "Sidi Ali Reis," *EI²*; Orhonlu, "Seydi Ali Reis," 39–56; Casale, *The Ottoman Age of Exploration*, 85–9.

17 On the historical literature and translations of the *Book of the Ocean*, see Chapter 1.

18 Casale, *The Ottoman Age of Exploration*, 186–7.

19 On the empirical methods used by these navigators to determine latitudes, see Alfred Clark, "Medieval Arab Navigation on the Indian Ocean: Latitude Determinations," *Journal of the American Oriental Society* 113, no. 3 (1993): 360–73.

20 Seydi Ali Reis, *Kitab-ı Muhit*, Süleymaniye Library, Nuruosmaniye 2948, 2a.

21 In Ottoman Turkish, *muhit* means "ocean." In Arabic, *muhit* is the active participle of the verbal noun *ihata*, meaning that which surrounds.

22 Seydi Ali Reis, *Kitab-ı Muhit*, Süleymaniye Library, Nuruosmaniye 2948, 2b.

At first glance, the *Book of the Ocean* appears to be a standard and, at times, tedious navigation manual full of technical details that might bore a reader outside the sailing profession. It does not contain any maps.[23] The value of the work, however, stems from its comprehensive treatment of the topography and climate of the Indian Ocean. In 10 chapters, Seydi Ali Reis deals with a variety of topics, such as the history of Portuguese discoveries in the region; the names and positions of the stars and the division of the firmament; common curses used by the mariners of India, Arabia, and Sri Lanka; the islands and atolls in the Indian Ocean; and the strong winds and storms of the region.[24] Seydi Ali Reis's thorough treatment of the islands, atolls, and ports around the Indian Ocean, his warnings about the possible dangers awaiting seamen in the region, and his tips on how to avoid and survive these calamities are clear indications of his experience and his knowledge of geographical literature on the Indian Ocean.[25] In this respect, Seydi Ali Reis follows in the footsteps of Piri Reis. Like Piri Reis, Seydi Ali Reis also incorporates his own experiences in the region into his account and he tries to transmit the latest knowledge about the region to his readers. Therefore, the *Book of the Ocean* is more than a sailor's manual. In it, Seydi Ali Reis refers to current events and relates the history of Portuguese discoveries in the Indian Ocean. In the fifth chapter, he describes how Portuguese discoveries started in the sixteenth century, which routes early Portuguese explorers used, and which islands they discovered first. Here, he states that he heard some of the details of the discoveries from a Portuguese seaman who entered the service of Süleyman during the grand vizierate of İbrahim Pasha.[26]

Seydi Ali Reis, like Piri Reis, wrote the *Book of the Ocean* in hopes of enhancing the geographical knowledge of the Ottoman court and at the same time shaping its political agenda. Seydi Ali Reis wrote under the patronage of Grand Vizier Rüstem Pasha, whose political and economic policies were much more conservative than those of his predecessor İbrahim Pasha. Rüstem Pasha, from the outset, was more concerned with consolidating the Ottoman authority over the areas that the empire was already in control of than adding new lands to the imperial domains. The *Book of the Ocean* articulated the grand vizier's political agenda and relegated the Indian Ocean to the margins of the Ottomans' imperial vision.

The main theme of Seydi Ali Reis's *Book of the Ocean* is the calamities one might experience in the region. On several occasions, Seydi Ali Reis reinforces this point by comparing the geographical size, climate, navigational techniques, and methods required when sailing in the Indian Ocean with the size, climate, and techniques needed in the Mediterranean. In a passage where he describes how a compass works, for instance, he discusses the different compasses used in France, Portugal, and the Ottoman Empire. Here, Seydi Ali Reis maintains that a broken

23 This might account for the scant amount of manuscript copies produced.

24 Seydi Ali Reis, *Kitab-ı Muhit*, Süleymaniye Library, Nuruosmaniye 2948, 4a–9b; 14a–15b.

25 Ibid., 15b–16a; 31b–32b.

26 Ibid., 28a–29a.

compass is not likely to cause too many problems for seamen in the Black Sea or the Mediterranean, where the distances are much shorter than in the Indian Ocean.[27] In making these comparisons, Seydi Ali Reis clearly wants to draw the court's attention to his professional knowledge about the Indian Ocean and the Mediterranean. But in doing so, he presents a rather gloomy picture of the Indian Ocean in comparison to the Mediterranean.

Although modern historical literature considers it an example of the Ottomans' imperial interest in the Indian Ocean, the *Book of the Ocean* did not find a large audience in the sixteenth century. Among the four extant manuscript copies of the work, only two date from the sixteenth century. That only two copies of the *Book of the Ocean* were produced in the sixteenth century while Piri Reis's *Book of Sea Lore* appeared in more than two dozen copies at that time suggests that geographical works on the Indian Ocean were not really of interest to the Ottoman court in the sixteenth century. Unlike the Mediterranean, where vital political and commercial interests were at stake, the Indian Ocean was of secondary importance for the majority of the members of the Ottoman court and their understanding of universal sovereignty. In the context of a "major concern for stability" in the second half of the sixteenth century,[28] Ottoman ruling elites such as Rüstem Pasha chose to leave the Indian Ocean aside and consolidate the energies of the empire into strengthening Ottoman sovereignty in the already conquered territories.

The *Book of the Ocean* is, however, invaluable as an example of the Ottoman court's openness to knowledge of the geography and history of the Indian Ocean. Seydi Ali Reis further delves into the geography and history of this region in his second book, the *Mirror of Lands*. In 1557, the admiral composed an account of his travels from the port of Surat to Istanbul for his prestigious patron, Rüstem Pasha.[29] Like the *Book of the Ocean*, the *Mirror of Lands* found its way into the palace library. There are five extant copies of the work, only two of which are from the sixteenth century.

The *Mirror of Lands* opens with an eloquent explanation of why the author composed this work. Seydi Ali Reis states that he prepared this account for posterity with the hope that it would one day appreciate the hardships he had endured on his journey. He describes how he was appointed admiral and given imperial orders to take the Ottoman naval ships abandoned in Basra to Egypt. He provides a detailed description of the difficulties that he and his crew faced at the hands of the Portuguese around the Gulf of Basra and the southern tip of the Arabian Peninsula. He also writes of the harsh and volatile weather conditions in the region. He tells of how he and his crew eventually navigated westward across the Indian Ocean until they finally anchored in the port of Surat three months after starting their journey in Basra.[30] In the following section, Seydi Ali Reis

27 Ibid., 16a–17a.

28 Tezcan, "The 'Frank' in the Ottoman Eye of 1583," 268–96.

29 Casale, *The Ottoman Age of Exploration*, 120–23.

30 Seydi Ali Reis, *Mir'atü'l-Memalik*, TKS. Revan Köşkü 1470, 2b–18b.

describes the geography of the Gujarat region, and places special emphasis on local rulers and on local Muslim shrines and communities. More than once he mentions the various Portuguese activities in the Red Sea and the Indian Ocean. But as its title suggests, the *Mirror of Lands* is about many lands, all of which were on the periphery of the Ottoman Empire. While India and the Indian Ocean receive the bulk of his attention, Seydi Ali Reis also provides descriptions of Transoxiana, Khurasan, Persia, and Iraq.

Textually speaking, the *Mirror of Lands* breaks from contemporary portrayals of Indian ports and cities. Whereas contemporary Portuguese accounts describe India as a land of endless riches and trade opportunities,[31] the *Mirror of Lands* projects India through the lens of Seydi Ali Reis's patron, Rüstem Pasha's political agenda. Here, Seydi Ali Reis uses India as a screen on which the new universal rulers—the Ottoman sultans—could project their imperial power and confirm their religious authority. Seydi Ali Reis frequently refers to the loyalty of Muslim communities in Gujarat to the Ottoman sultan, and to their desire to be part of the Ottoman Empire:

> Since the time of Adam, no corsairs, in other words no captain who is knowledgeable about the science of navigation, reached these parts of India from the Ottoman lands [*diyar-ı Rum*]. So God willing, one wishes that in the near future, Gujarat will be annexed to the Ottoman Empire [*Memalik-i Osmaniyye*]. Thus, these ports of India will be saved from the hands of the despicable infidels.[32]

It is important to note here that Seydi Ali Reis, perhaps in an effort to encourage the sultan's military engagement in the region, articulates the wishes of the local Muslims for the Ottomans to annex these lands. Although this passage can be considered as Seydi Ali Reis's plea to the sultan and to the grand vizier to take a more aggressive stance in India and the Indian Ocean to protect the Muslim communities, in the end, both seemed to take a more peaceful approach.[33] This is the only passage in the whole work where the author courageously tries to convince the Ottoman court to take action. In the rest of the work, the main emphasis is on the loyalty of the Muslim communities to the Ottoman sultan.

---

31 For an example of sixteenth-century Portuguese descriptions of the riches of the west coast of India, see Tomé Pires, *The Suma Oriental of Tome Pires: An Account of the East, from the Red Sea to China, Written in Malacca and India in 1512–1515* and *The Book of Francisco Rodrigues: Pilot-Major of the Armada that Discovered Banda and the Moluccas*, trans. and ed. Armando Cortesão (New Delhi: Asian Educational Services, 1990), vol. 1, 52–84. For Tomé Pires's description of the Indian Archipelago's riches, see Tomé Pires, *The Suma Oriental of Tome Pires*, vol. 1, 135–223.

32 Seydi Ali Reis, *Mir'atü'l-Memalik*, TKS. Revan Köşkü 1470, 19b.

33 On the *Mirror of Lands*'s importance for the Ottoman aspirations in the Indian Ocean, and its articulation of Rüstem Pasha's Ottoman-centric world view, see Casale, *The Ottoman Age of Exploration*, 84–7.

For instance, in one passage, Seydi Ali narrates a conversation between the Sultan of Gujarat's grand vizier and the Portuguese ambassador in which the Portuguese ambassador demands the return of Seydi Ali Reis and his crew to the Portuguese. The admiral claims that the grand vizier refused this request with the following words: "We need the Ottoman sultan [*Padishah-ı Rum*]. Had our ships not reached their trading ports, we would be ruined [our position would be altered]. Above all, he is the Sultan of Islam. Is it reasonable to demand his naval captain from us?"[34]

Such expressions of devotion and respect for the Ottoman sultan by Gujarat rulers and natives did not reflect reality. Ottoman-Gujarati relations had followed an uneven path since their first interactions in the early sixteenth century. Neither party trusted the other, and economic alliances with the Portuguese were not uncommon in Gujarati and Ottoman courts. Why, then, did Seydi Ali Reis insist on portraying Muslim communities in India as loyal to the Ottoman sultan? Seydi Ali Reis's mission in the Indian Ocean coincided with the second half of Sultan Süleyman's reign, when a distinct Ottoman imperial culture emerged in Constantinople.[35] During the first two decades of his reign, Sultan Süleyman's direct orders, military campaigns, and edicts determined the Ottoman state's affairs. By contrast, during the second half of his reign, the dissemination of dynastic law and the dispensation of justice became the symbols of the Ottoman sultan's impersonal authority.[36] This newly injected Sunni piety came in direct response to the Ottomans' rivalry with the Shi'i Safavids, and was meant to reinforce the dynastic legitimacy and spiritual authority of the Ottoman house.[37]

Seydi Ali Reis, who acted as an official ambassador on behalf of the Ottoman court to the Gujarat sultan and was the protégé of the Ottoman grand vizier,[38] contributed to this enterprise by emphasizing the Gujarati Muslims' respect for the Ottoman sultan and the warm welcome he and his crew received from Gujarati elites. Seydi Ali Reis's message here is clear: the Ottoman sultan's claims to universal sovereignty were well received and accepted even in India. In the *Mirror of Lands*, Seydi Ali further strengthens this point in several anecdotes about his conversations with Gujarati elites on his way from Surat to Constantinople. In these passages, Seydi Ali Reis insists that the Ottoman Empire and its capital city are unrivaled in splendor and beauty. During a conversation with the Gujarati ruler, Seydi Ali Reis even compares the Ottoman lands to India and affirms that the size of the Ottoman Empire equals that of Alexander the Great's—both, he says,

---

34 Seydi Ali Reis, *Mir'atü'l-Memalik*, TKS. Revan Köşkü 1470, 23b.

35 Necipoğlu, "Süleyman the Magnificent and the Representation of Power," 401–27; idem, "A Kanun for the State, a Canon for the Arts"; Fleischer, "The Lawgiver as Messiah."

36 Fleischer, "The Lawgiver as Messiah," 166; idem, *Bureaucrat and Intellectual*, 191–252.

37 Fleischer, "The Lawgiver as Messiah," 161.

38 Suraiyya Faroqhi argues that Seydi Ali Reis wished to present himself as an ambassador. He fulfilled some of the duties of an ambassador and took back useful information to Constantinople. Faroqhi, *The Ottoman Empire and the World Around It*, 184.

have possessions in the seven climes of the world.[39] Later on, in describing his visit to the Safavid shah's court, Seydi Ali Reis maintains that he boldly defended his capital against the shah:

> One day, the shah asked this unworthy servant: Which city among the ones that you have seen around the world impressed you the most? This humble servant answered with this poem: "Although I visited every city in this world, I never saw anything comparable to Istanbul and Galata."[40]

In the *Mirror of Lands*, Seydi Ali Reis presents himself as the proud representative and defender of the Ottomans' imperial project. He uses India as a mirror to reflect the Ottomans' grandeur and imperial image. In Seydi Ali Reis's account, Portuguese traveler Tomé Pires's (d. 1540) description of the riches and trade opportunities in the region are conspicuously absent. Seydi Ali Reis does not even refer to the importance of Surat for trade, for example.[41] We learn nothing of the physical appearance of Indians, and nothing about their language.[42] He frequently refers to the Muslim communities in the region and their loyalty to the Ottoman sultan. In making these references, he lends support to the Ottomans' imperial claims, while at the same time carefully placing India on the periphery of the Ottoman world.

With the Ottoman court's attention fixed, for the most part, on the Mediterranean, both the *Book of the Ocean* and the *Mirror of Lands* went almost unread in the sixteenth century. The *portolan* atlases produced in Europe for the Ottoman market also reflect and confirm this comparatively limited interest in the region. For example, the *Walters Sea Atlas* (1560), which charts the Mediterranean in seven different *portolan*s, features only a single map of the Indian Ocean (Figure 3.3).[43] The other *portolan* atlases prepared for the Ottoman court during this period either completely omit the region, as in the case of Ali Macar Reis's atlas, or depict only the Arabian Sea, as in the case of the *Imperial Atlas*. Nevertheless, the Ottoman court's marginal but enduring intellectual and

---

39 Seydi Ali Reis, *Mir'atü'l-Memalik*, TKS. Revan Köşkü 1470, 39b–41a.

40 Ibid., 75a. In the same passage, the author again refers to the Ottoman Empire as the *Diyar-ı Rum* (Roman domains).

41 The Port of Surat rose in importance during this period when Diu was under Portuguese control. See Subrahmanyam, "A Note on the Rise of Surat in the Sixteenth Century."

42 On Portuguese views of the Indians in the sixteenth and seventeenth centuries, and on aspects of "similarities" in Portuguese accounts, see Michael Adas, "*Machines as the Measure of Men: Science, Technology and Ideologies of Western Dominance* (Ithaca: Cornell University Press, 1989); Michael N. Pearson, "The Search for Similar: Early Contacts between Portuguese and Indians," in *Clash of Cultures: Essays in Honour of Niels Steensgaard*, ed. Jens Christian V. Johansen, Erling Ladewig Petersen, and Henrik Stevensborg (Odense: Odense University Press, 1992), 144–59.

43 The map outlines recent European discoveries in the area. Thomas Goodrich, "The Earliest Ottoman Maritime Atlas: The Walters Deniz Atlası," 49–50.

political interest in India and the Indian Ocean continued well into the reign of Murad III (r. 1574–95), a period that saw a surge in the number of manuscripts produced in the Ottoman Empire. As the amount of universal histories and chronicles, decorated primarily with miniature paintings, increased under the patronage of the Ottoman court, Ottoman geographers also contributed to this flourishing intellectual and artistic environment.

As the second and third chapters have shown, geographical works on the Mediterranean and Constantinople circulated widely in the Ottoman court. A significant example of how Ottoman intellectuals answered the Ottoman court's interest in the history and geography of far-off lands during this period is Seyfi Çelebi's (d. ca. 1590) *Book on the Histories of the Rulers of India and China.*[44] In nine chapters, Seyfi Çelebi, an Ottoman bureaucrat, depicts the history, geography, economic life, customs, and rulers of China, India, Transoxiana, and Persia. This work is detailed and thorough yet markedly less ornamented in comparison to similar works of the period. So far, several historians have tried unsuccessfully to identify the author.[45] In the opening lines of his work, Seyfi Çelebi describes himself as a provincial treasurer who compiled his work on behalf of the reigning sultan, Murad III.[46]

Written during the peak of the Ottoman-Safavid conflict, Seyfi Çelebi's work, together with the other historical and geographical works from the same period, helped to satisfy the Ottomans' curiosity about the geographies and societies of different regions, while also reinforcing the intellectual effort to articulate Ottoman imperial claims at the end of the sixteenth century. Seyfi Çelebi opens his account by explaining why he wrote the book: "history is such a noble learning [...] that if the rulers learn about the previous rulers and how they ruled, they will become better rulers. Also, if they learn that all these previous rulers [eventually] passed away, they will not get so attached to worldly affairs and will be more concerned with the adulation of God."[47] The first four chapters of the work discuss the rulers of China, Transoxiana, and Persia. In the fifth chapter, Seyfi Çelebi turns his focus to India and its rulers. In the introduction of this chapter, he provides a brief description of the geography of India: "India is a very vast land. Its greatness is recorded in the histories. The land is completely populated. There aren't many deserts and its forests are flat. Even its mountains are scarce. Some of the produce grows twice in India."[48] After this very short account on India's geography, Seyfi Çelebi delves

---

[44] There are only two extant manuscripts of this work, one in Leiden and one in Paris. Seyfi Çelebi, *Kitab-ı Tevarih-i Padishahan-ı Vilayet-i Hindu ve Hitay*. Leiden University Library Cod. 917 (1) Warn.; Paris National Library Suppl. Turc. No. 1136. Joseph Matuz prepared the facsimile edition of the Leiden copy in 1968: Joseph Matuz, ed., *L'Ouvrage de Seyfi Çelebi: Historien Ottoman du XVIe Siecle* (Paris: Dépositaire, 1968).

[45] Franz Babinger, *Die Geschichtsschreiber der Osmanen und Ihre Werke* (Leipzig: Otto Harrassowitz, 1927), 69; Matuz, ed., *L'Ouvrage de Seyfi Çelebi*; Christine Woodhead, "Seyfi," *EI*².

[46] Matuz, ed., *L'Ouvrage de Seyfi Çelebi*, 163.

[47] Ibid.

[48] Ibid., 186.

into the history of India's rulers and how they came to rule the country.[49] Although there is no evidence that Seyfi Çelebi visited the places featured in his book, the anecdotal style of his work suggests that he had been in contact with travelers and merchants from these areas. His historical account does not add anything new to the Ottomans' existing geographical or historical knowledge about India, but the fact that Seyfi Çelebi reserves a chapter on India in his work suggests that for the Ottoman court India was still an important and valued region, albeit peripheral to the Ottoman world.

From the early decades of the sixteenth century, the Ottomans had been aware of India and its role in trade and politics in the Indian Ocean. Ottoman political and intellectual interest in India and the Indian Ocean, however, remained subdued. The Ottoman court in the end acknowledged the limits of their military power in this area. While a full-scale military campaign to secure the region was never feasible, the Ottoman sultan's image as the guardian of Islam provided Ottoman geographers with the knowledge needed to fold India into the empire, even if it was only to remain on the periphery.

The New World: Ottoman Peripheries to the West

Another part of the globe where Ottoman imperial conquest was not possible was the New World. As early as the fourteenth century, Genoese and Portuguese seamen discovered new territories in the Atlantic. By the beginning of the fifteenth century, European sailors reached the Canary Islands, Madeira, and the Azores. At first, these discoveries did not change Europe's perception and understanding of the geography of the world, nor did they challenge Ptolemy's tripartite division of the world. Reports of these voyages were incomplete and contradictory. Cartographers and cosmographers were unsure about how to integrate these new territories into their maps. Many geographers and intellectuals were still confident that the world consisted of three main continents: Asia, Europe, and Africa. The newly discovered territories were long considered part of Southeast Asia.[50] Columbus's discoveries in the Caribbean, and the Labrador coast, the Brazilian coast, and Newfoundland (discovered by John Cabot, Pedro Álvares Cabral, and Cortes Real) were initially recorded as islands in the Ocean Sea. The earliest surviving terrestrial globe (from 1492), created by Martin Behaim, illustrated these new discoveries as a belt of islands between Europe and Southeast Asia.[51]

---

[49] After the chapter on India, in the rest of the work Seyfi Çelebi focuses his attention mainly on the Sunni Muslim rulers of Central Asia. Seyfi Çelebi was obviously interested in telling the Ottoman sultan the history of Central Asia. He also clearly aspired to impress the sultan with his knowledge of the history of Muslim rule in Central Asia in order to earn the sultan's patronage. As shown in the first chapter, the increasing focus in this period on the Ottoman lineage going back to Central Asia provided the perfect context for Seyfi Çelebi's history. Ibid., 188–94.

[50] Brotton, *Trading Territories*, 26–35.

[51] Cosgrove, *Apollo's Eye*, 82–3.

At the beginning of the sixteenth century, as more lands were discovered, it became increasingly difficult to depict these territories on maps. In 1507, Martin Waldseemüller wrote: "The earth is now known to be divided into four parts. The first three parts are continents while the fourth part is an island, inasmuch as it is found to be surrounded on all sides by the ocean."[52] Nevertheless, in the world map (Figure 4.1) that accompanied the text, Waldseemüller depicted the lands discovered by the Portuguese and the Spanish as a single continent—separate from Asia—on the far side of the Atlantic.[53]

Even after Magellan's 1520–22 circumnavigation, European cartographers did not fully embrace the idea of a separate continent; for many, the newly discovered territories remained part of Asia. The De Bure globe, for instance, which was originally created around 1528, still depicted the Americas as a peninsula of China.[54] Only after Vitus Bering's second expedition in 1741 did America's physical separation from Eurasia become accepted knowledge.[55]

Cartographers struggled to integrate the new territories into their maps. But this was hardly the only problem caused by the discoveries. The expeditions in the Atlantic Ocean and the discovery of the New World also led to territorial and economic rivalries between the Iberian monarchies of Portugal and Spain. In 1481 and 1494, two papal declarations tried to resolve the intra-Iberian rivalry over these so-called "islands" in the Atlantic Ocean. The conquests of Mexico and Peru, however, challenged this notion of an insular belt. As the Indies' political and economic importance grew, it became possible to talk about an "empire" in the New World. Both cartographers (from the *Casa de la Contratación*) and historians gave shape and meaning to this dream.[56] Gonzalo Fernandez de Oviedo's *Sumario de la natural historia de las Indias* (1526) and *Historia general y natural de las Indias* (1530), and Pietro Martire d'Anghiera's *De orbe novo decades* (1530) not only related stories of the new territories and peoples but also offered maps of the Indies as a whole. These works were later translated into Italian. *De Orbe*, for example, was published in Italian in Venice in 1534.[57]

The Ottoman court and Ottoman geographers actively participated in these geographical and political discussions about the New World. It has long been suggested that the Ottomans had very little contact with the New World in the sixteenth century. Since the New World was geographically very distant from the Ottoman domains, it did not attract the attention of the court or Ottoman intellectuals. Hardly anyone from the empire traveled to the New World, and none

---

52 Martin Waldseemüller, *Cosmographiae introductio cum quibus dam geometrieae ac astronomiae principiis ad eam rem necessariis* (St. Die, France, 1507; n.p.: Readex Microprint, 1966), 70; cf. Cosgrove, *Apollo's Eye*, 83.

53 Padrón, *The Spacious World*, 21–3.

54 Cosgrove, *Apollo's Eye*, 83.

55 Ibid.

56 Padrón, *The Spacious World*, 138.

57 The translation included a more comprehensive map, which was most likely derived from the cartographic resources of the *Casa de la Contratación*. Padrón, *The Spacious World*, 138–41.

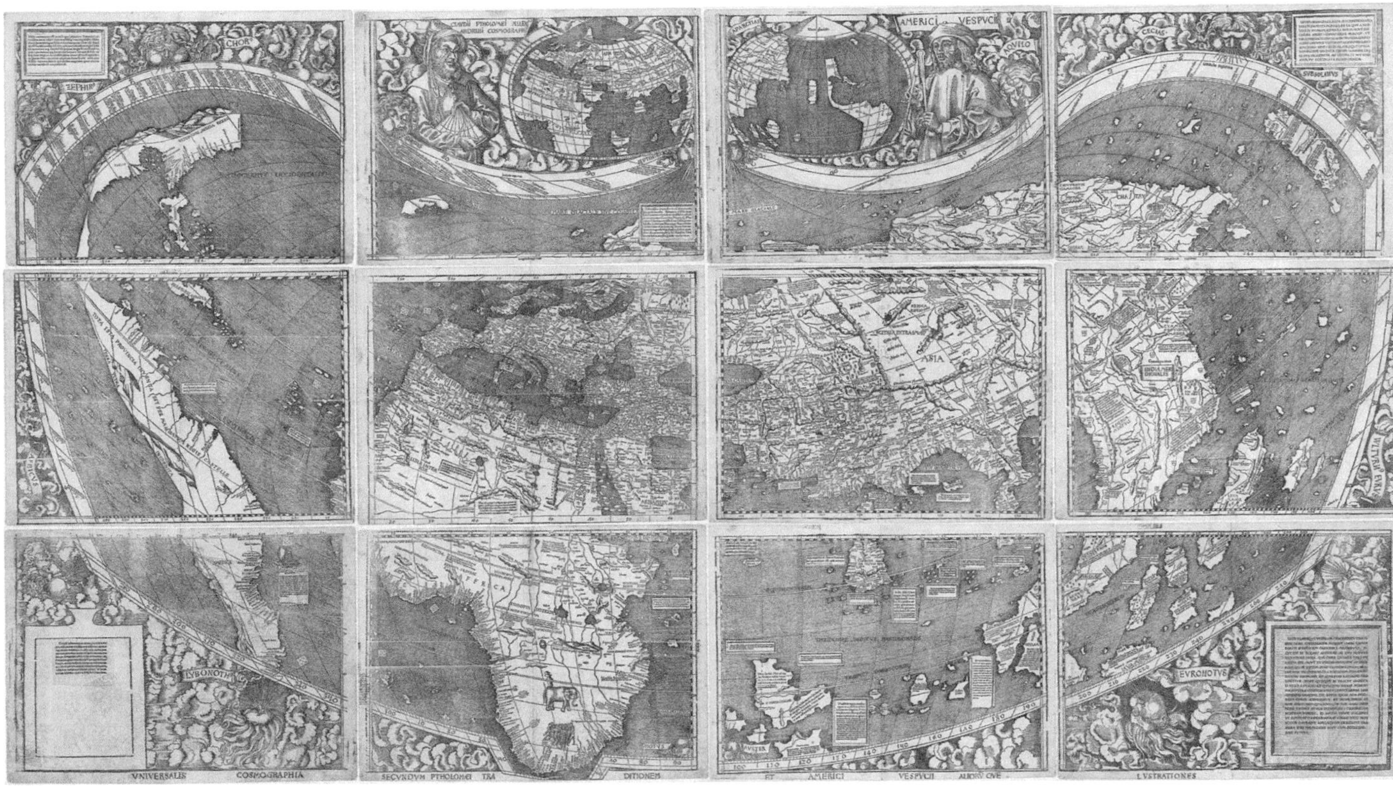

Fig. 4.1 Martin Waldseemüller World Map (1507). Courtesy of the Geography and Map Division, Library of Congress.

left accounts of their journey.[58] This view, however, is very deterministic. It does not take into account that, despite its geographical separation, the New World and its resources had an impact on Ottoman society, politics, and geographical understanding in the sixteenth century. The new agricultural products entering the European market—including tobacco, maize, and tomatoes—eventually found their way to the Ottomans' lips. The influx of American bullion into Europe also damaged the Ottoman economy and led to a devaluation of coinage at the end of the sixteenth century.[59] Interestingly, the first colonizers dubbed the *Meleagris gallopavo* they encountered in North America "turkey."[60]

The Ottoman court and its ruling elites were not as oblivious to the new discoveries as the canonical literature might lead one to believe. In the sixteenth century, major imperial rivalries were still alive in the Mediterranean. The Ottoman Empire, just like its Spanish counterpart, focused its energies and capital on maintaining order in the Mediterranean. It is not surprising, then, that the majority of the sea charts prepared by the Spanish *Casa de la Contratación* were still *portolan* charts centered on the Mediterranean.[61] Like their Spanish colleagues, Ottoman geographers and historians showed increasing interest in the New World and the discoveries made in the Atlantic Ocean. The earliest example of this endeavor is a world map created in 1513 by none other than Piri Reis, who presented the map to Selim I in 1517 (Figure 4.2).[62]

---

[58] G. Guevara Bazan, "Muslim Immigration to Spanish America," *The Muslim World* 56 (1966): 173–87; G. G. Bazan, "Some Notes for a History of the Relations between Latin America, the Arabs and Islam," *Muslim World* 61 (1971): 284–92; Thomas Goodrich, *The Ottoman Turks and the New World: A Study of Tarih-i Hind-i Garbi and Sixteenth-Century Ottoman Americana* (Wiesbaden: Otto Harrassowitz, 1990), 7–9.

[59] Ömer Lütfi Barkan, "The Price Revolution of the Sixteenth Century," *International Journal of Middle East Studies* 6 (1975): 3–28.

[60] They named the bird "turkey" because of its similarity to the "turkey fowl," which was at the time imported to Europe via the Ottoman lands.

[61] Alison Sandman, "Spanish Nautical Geography in the Renaissance," in *The History of Cartography, Vol. 3, Bk. 1, Cartography in the European Renaissance*, ed. J. B. Harley and David Woodward (Chicago: University of Chicago Press, 2007), 1096. In addition to works generated at the *Casa*, charts of the Mediterranean continued to be produced by Catalan cartographers. In different parts of Europe, small family workshops continued to produce *portolan* charts and atlases that focused exclusively on the Mediterranean. Corradino Astengo, "The Renaissance Chart Tradition in the Mediterranean," in *The History of Cartography, Vol. 3, Bk. 1, Cartography in the European Renaissance*, ed. J. B. Harley and David Woodward (Chicago: University of Chicago Press, 2007), 174–5.

[62] For a facsimile and for English, French, and German translations of Piri Reis's world map, see *Piri Reis Haritası Hakkında İzahname*, ed. Yusuf Akçura (Ankara: Türk Tarih Kurumu, 1992). On Piri Reis's world map, see Svat Soucek, *Piri Reis and Turkish Mapmaking after Columbus* (Oxford: Oxford University Press, 1996). On the connection between Piri Reis and Selim I's interest in the New World, see Andrew Hess, "Piri Reis and the Ottoman Responses to the Voyages of Discovery," *Terrae Incognitae* 6 (1974): 19–37. On the connection between Piri Reis's world map and Ottoman interest in the Indian Ocean, see Giancarlo Casale, *The Ottoman Age of Exploration*, 23–5.

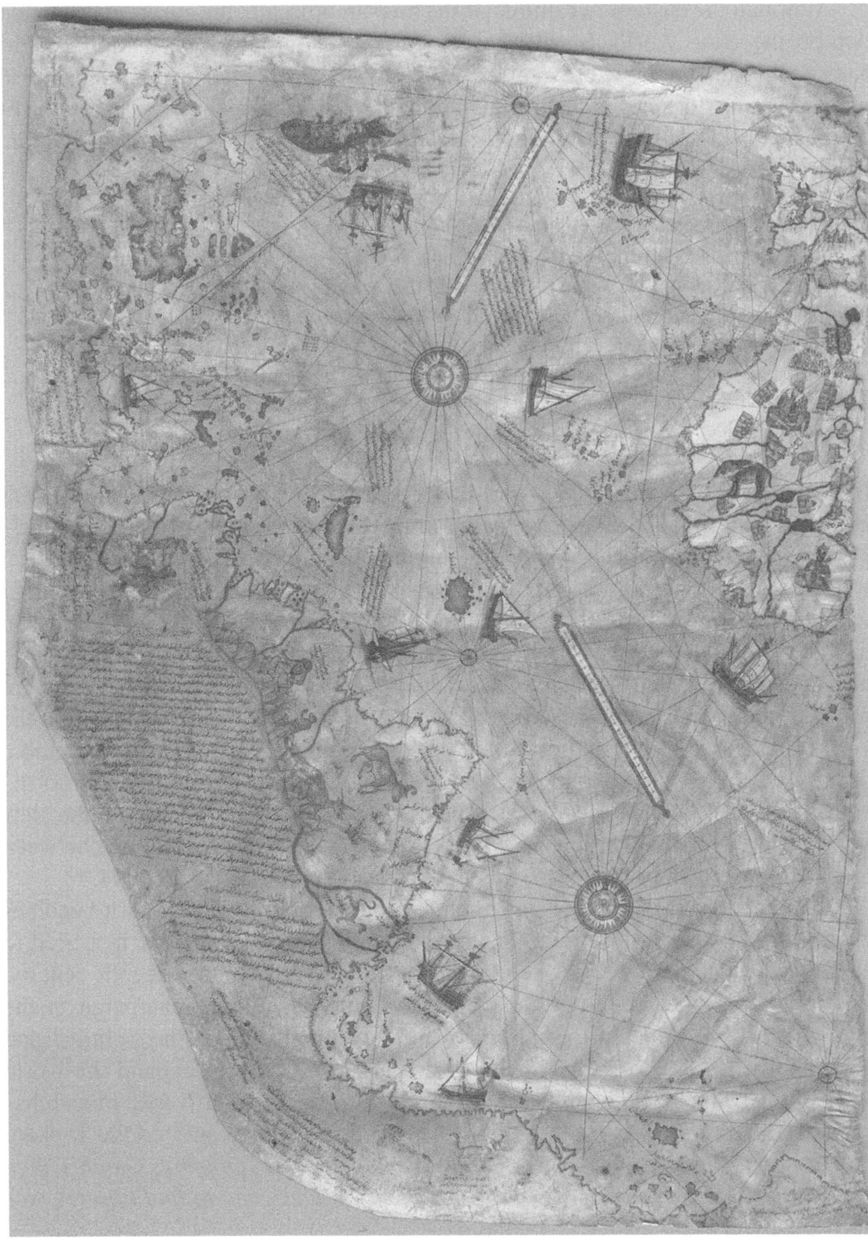

Fig. 4.2 Piri Reis World Map (1513). Piri Reis, *Mappamundi*, TKS. Revan Köşkü 1633 mük. The Library of the Topkapı Palace Museum.

Although the eastern two-thirds of the map are missing, the surviving segment depicts the New World. Similar to contemporary European cartographers, Piri Reis was unsure about where to place the newly discovered territories on his world map. The map does not indicate whether the new discoveries were part of Asia or an independent continent. In one of the legends ornamenting the Atlantic Ocean, Piri Reis says:

> This sea is called the Western Sea, but European sailors call it the Mare d'Espagna, which means the Sea of Spain. Up to now it was known by these names, but Colombo, who has opened up this sea and made these islands known, and also the Portuguese infidels who opened up the region of India, have agreed together to give this sea a new name. They have given it the name Ovo Sano (Ocean), that is to say, sound egg. Before this, it was thought that the sea had no end or limit and that at its other end was darkness. Now they have seen that this sea is girded by a coast; because it is like a lake, they have called it Ovo Sano.[63]

In the legend to the left of his world map, Piri Reis says that he drew on different European and Muslim maps circulating in the Mediterranean to create this work. Here, he also delves into the history of Columbus's discoveries in the New World. Piri Reis says that he learned about these discoveries, about Columbus's first encounter with the natives, and about the treasures of the region from a Spanish slave who had been with Columbus during his first three ventures.[64] The introductory lines of this part read: "This section tells how these shores and also these islands were found. These coasts are named the shores of Antilia. They were discovered in the year 1491. But it is reported thus that a Genoese infidel, his name was Colombo, discovered these places."[65] In this section, Piri Reis also draws attention to the valuable goods—such as pearls and gold—that early colonizers found in the New World. Moreover, another legend on the Atlantic Ocean refers to the line between Spain and Portugal that was drawn with the Treaty of Tordesillas in 1494.[66]

When analyzed with other geographical sources from this period, it becomes clear that Piri Reis's world map is an important example of a work that tried to enhance the Ottoman court's geographical knowledge in the sixteenth century. Piri Reis does not encourage the sultan to launch a military campaign in the New World, but rather informs him of new developments in the political and geographical structure of the world. In doing so, he seeks to expand the world view of the sultans who claimed to be world conquerors and universal monarchs.

In 1560, almost five decades after Piri Reis's world map entered the Topkapı Palace library, another world map, entitled *Kemaliyle Nakş Olunmuş Cümle Cihan Numunesi* (*The Representation of the Whole World Designed in Its Entirety*), was

---

63 *Piri Reis Haritası Hakkında İzahname*, 12.

64 Piri Reis, *Mappamundi*, TKS. Revan Köşkü 1633 mük.

65 *Piri Reis Haritası Hakkında İzahname*, 12. Here, Piri Reis dates the discoveries to 1491, still one year earlier than their actual date.

66 Piri Reis, *Mappamundi*, TKS. Revan Köşkü 1633 mük.

prepared for the Ottoman court (Figure 4.3). According to the text encircling the map, a certain Tunuslu Hacı Ahmed drew this heart-shaped geographical projection on six conjoined wooden blocks. The blocks were later found in the archives of the Venetian Council of Ten in Venice in 1794. We know of 24 copies that were made from these wooden blocks.[67] Today, several copies survive in libraries throughout Europe.[68]

The origin of the Hacı Ahmed World Map is still disputed. However, what little we know confirms that Marc Antonio Giustinian—a Venetian printer—prepared the map in a workshop in Venice with the Ottoman market in mind, and that Hacı Ahmed was a fictitious character.[69] As for the audience of the map, there is not enough evidence to prove whether or not Hacı Ahmed's world map reached the Ottoman capital in the sixteenth century. In 1568, Giustinian was called before Venetian authorities who revoked his license to print the map and confiscated the wood blocks. However, circumstantial evidence suggests that Hacı Ahmed's *mappamundi* might have made it into the hands of one of Süleyman the Magnificent's three sons, all of whom were interested in Venetian cartographical works as symbols of prestige and power during their struggle for the throne.[70]

Hacı Ahmed's world map has so far been regarded as one of the most impressive examples of cultural and political exchange between the Ottoman Empire and Europe.[71] For our purposes, however, this is only a small part of

---

[67] Marie Armand Pascal d'Avezac, "Note sur une mappemonde turke du XVIe siècle conservée à la Bibliothèque de Saint-Marc à Venise," in *Reprint of Studies on the Ottoman Cartographers Piri Reis (d. 1554) and Haggi Ahmad (d. about 1560)*, vol. 22 of *Islamic Geography*, ed. Fuat Sezgin et al. (Frankfurt : IGAIW, 1992), 315–99.

[68] Haci Ahmed, *Mappamundi*, Bodleian, March, nr. 454; Paris Bibliotheque Nationale 1865. There are also two copies in St. Mark's library, one in Vienna, one in the library of the seminary of St. Maria della Salute, and one in the Museo Correr. The Biblioteca Marciana in Venice, where the original wooden blocks are kept, also holds a copy dating to the nineteenth century. V. L. Ménage, "The Map of Hajji Ahmed and its Makers," *BSOAS* 21 (1958): 293. In addition, there are also copies at the John Carter Brown library at Brown University, the Clements Library at the University of Michigan, the Newberry Library in Chicago, and the Library of Congress in Washington, DC.

[69] Ibid., 299–312. Ménage concludes that "the compiler of the text is a non-Muslim with imperfect knowledge of both Arabic and Ottoman Turkish and a good but not intimate knowledge of the Muslim World and Muslim habits of thought, someone who was acquainted with Venetian cosmographer Giovanni Battista Ramusio's works and someone who was familiar with the language of Ottoman chancery and who drew up the companion text in Latin script." See also Jerry Brotton, "Printing the World," 37.

[70] Benjamin Arbel demonstrated that Selim—the future Sultan Selim II—ordered world maps from the workshops in Venice in the early 1550s, shortly before the struggle for the throne began. Benjamin Arbel, "Maps of the World for Ottoman Princes? Further Evidence and Questions Concerning 'The Mappamundo of Hajji Ahmed,'" *Imago Mundi* 54 (2002): 21–30; Antonio Fabris also showed that Bayezid, another son of Süleyman I, was interested in world maps prepared in the workshops of Venice. Antonio Fabris, "The Ottoman Mappa Mundi of Hajji Ahmed of Tunis," *Arab Historical Review for Ottoman Studies* 7–8 (1993): 31–7.

[71] Ian Manners, *European Cartographers and the Ottoman World*, 21–3.

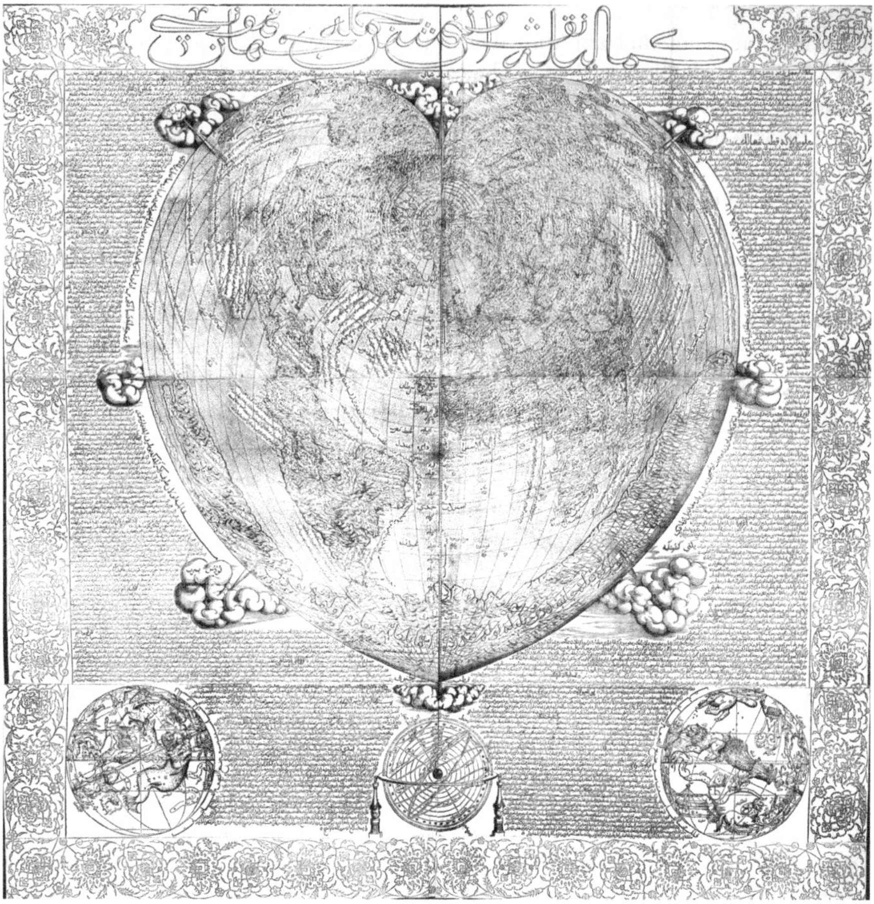

Fig. 4.3 Hacı Ahmed's World Map (ca. 1560). Hacı Ahmed, *Mappamundi*, Newberry Library, Novacco 8F 11. Courtesy of Newberry Library.

the map's significance. More importantly, Hacı Ahmed's world map represents a strong connection between the geographical and political discussions on the New World and the Ottoman court. It does point to the awareness in Europe of the openness of the Ottoman court to new developments in the geographical and political structure of the world.[72]

In addition to imparting current knowledge about the geography of the New World, the long text surrounding Hacı Ahmed's map describes the continents, the 12 principal countries of the world, and the world's seven rulers. Accompanying the text are over 20 legends that describe the spice trade and the Portuguese

[72] The map bears a striking resemblance to an earlier one designed by Oronce Fine—a French geographer—and first printed in 1534. Ibid.

control of trade between the Pacific and Indian oceans.[73] In the first 13 lines, the cartographer says that he prepared this map in order to introduce his patron, the Ottoman sultan, to the latest European cartographical productions, which were designed in accordance with the teachings of Plato, Socrates, Abu al-Fida, and Loqman. After giving a general and incomplete literary definition of latitude and longitude, the Equator, the Tropics, and the Poles, the author continues with an abridged description of the world's four continents: Asia, Africa, Europe, and the New World. Along with the names of the continents, the author provides the longitude and latitude of each one, and names each continent's principal countries and their rulers.[74] In the description of the 12 countries of the world and the world's seven rulers, the author pairs the lands and cities with fixed stars, and the states and their rulers with the zodiac and the planets. Thus, he creates an image of the rulers and their states illuminating the lands and cities.[75]

In this model, Maghrib corresponds to Cancer; Negrita and Abyssinia to Scorpio; Temixtitan (New Spain) to Virgo; Peru to Aquarius; Turkistan to Sagittarius; Arabia to Taurus; the Land of Monomotapa (the hinterland of Mozambique) to Gemini; Italy to Leo; Portugal to Pisces; Aleman (consisting of England, Scotland, Denmark, and Sweden) to Aries; Sarmatia (the lands between Aleman in the West and the Don in the East, and the Danube and the Black Sea in the South) to Capricorn; and, finally, Malabar to Libra. There are seven rulers dominating these lands and he assigns a planet to each: the Khan of Cathay and Khotan (the Moon), the Emperor of Abyssinia (Saturn), the French king (Venus), the Spanish king (Jupiter), the Safavid shah (Mars), and the Bengali emperor (Mercury). Each of these rulers revolves around the Ottoman sultan (the Sun). The text does not assign any star to the New World; however, it states that it has just been discovered and is now ruled by the Spanish king (Jupiter).[76]

In the early modern era, observations and measurements associated with the heliocentric hypothesis first put forward by Copernicus (d. 1543) and later improved by Johannes Kepler (d. 1630) challenged the central position of the world and instead located the sun at the center of the universe.[77] That the text of Hacı Ahmed's world map singles out the Ottoman sultan as the sun suggests that European mapmakers were aware of the sultan's imperial aspirations. Wanting to gain access to the Ottoman market, cartographers made use of this knowledge by incorporating cosmographical findings into their works to create images that would please potential Ottoman buyers. Earlier texts considered Alexander the Great the Sun. Now the sultan was in the center, illuminating Europe, Africa,

---

[73] George Kish, *The Suppressed Turkish Map of 1560* (Ann Arbor: William L. Clements Library, 1957), 11.

[74] Ménage, "The Map of Hajji Ahmed and Its Makers," 295–6.

[75] Ibid., 297.

[76] Hacı Ahmed, *Mappamundi*, Bodleian, March, nr. 454, lines between 97 on the right and 127 on the left.

[77] Thrower, *Maps and Civilization*, 91–100.

and Asia. In this arrangement, he was viewed as superior to other rulers, who were merely planets.[78] Therefore, Hacı Ahmed's world map fulfilled two purposes: on the one hand it articulated the Ottoman imperial claims for universal sovereignty with a sophisticated geographical scheme and introduced the recent geographical knowledge about the New World to the Ottoman court, on the other.

The Ottoman court's thirst for recent geographical and political knowledge about the New World also inspired Ottoman geographers to integrate as much information on the New World as possible into their works. For instance, Seydi Ali Reis opens his *Book of the Ocean* with a short description of the discoveries made in the Atlantic Ocean. In his introduction, Seydi Ali Reis states that although scholars had previously believed that only a quarter of the northern part of the earth was known, the Portuguese later found a town that was part of a continent situated west of the Canary Islands. The author explains that this continent is called the New World and that it can already be found on the *mappamundi*.[79] In the fourth chapter, Seydi Ali Reis informs his readers of the geographical coordinates of the New World.[80] After noting that the people of the New World do not subscribe to any of the three great religions and that they had not heard about the prophets,[81] he describes their physical appearance, and claims that the Portuguese tried unsuccessfully to transport two of these creatures to their capital city.[82] Seydi Ali Reis also makes a point of noting that the Portuguese, French, and Spanish had ventured to this new land and immediately exploited its resources.[83] Seydi Ali Reis's effort to inform the court of such news served to underscore his knowledge. At the same time, it suggests that he knew of the court's desire to learn of the Europeans' latest discoveries.

In the next decades, the Ottoman court continued to commission geographical and historical accounts on the New World. In the 1580s, an anonymous account titled *Tarih-i Hind-i Garbi* (*History of the India of the West*) entered the palace library in Constantinople.[84] Dedicated to Murad III, this work was the most comprehensive account available on the New World until the seventeenth century.[85]

---

78 Hacı Ahmed, *Mappamundi*, Bodleian, March, nr. 454, lines between 97 on the right and 127 on the left.

79 Seydi Ali Re'is, *Kitab-ı Muhit*, Süleymaniye Library, Nuruosmaniye 2948, 4a.

80 Ibid., 28a.

81 Ibid.

82 Ibid., 28b.

83 Ibid.

84 *Tarih-i Hind-i Garbi*, Beyazit Library, 4969; TKS. Revan Köşkü 1488. For an English translation and critical analysis of the work, see Goodrich, *The Ottoman Turks and the New World*; for a modern Turkish transcription and English translation of the text, see *Tarih-i Hindi-i Garbi veya Hadis-i Nev: History of the West Indies Known as the New Hadith*, ed. Tülay Duran, trans. and transcribed by Fuat Yavuz, Ayhan Özyurt, Tevfik Temelkuran (Istanbul: Tarihi Araştırmalar Vakfı, Istanbul Araştırma Merkezi, 1999).

85 Karamustafa, "Maps and Plans," 218. On Katip Çelebi's treatment of the New World and his sources, see Gottfried Hagen, "Katip Çelebi and Tarih-i Hind-i Garbi," *Güney-Doğu Avrupa Araştırmaları Dergisi* 12 (1998): 101–15.

So far, *History of the India of the West* has been valued for its inclusion of major contemporary European sources and as a symbol of the Ottomans' interest in the Indian Ocean.[86] For this study's purposes, this work represents the pinnacle of the Ottomans' geographical and political interest in the New World. The author of this work attempted to transmit the latest geographical and political knowledge about the region to the Ottoman court. He clearly believed that the Ottoman court should be conversant with the New World and aware of its geopolitical importance. Nevertheless, the court made use of a familiar template and located the New World on the peripheries of the Ottoman Empire, where imperial conquest was not feasible.

History of the India of the West is comprised of three chapters. The first chapter is a short cosmographic account of the universe. In the second chapter, the author describes the topography of the Mediterranean and the Atlantic Ocean. The last chapter focuses exclusively on the geography, topography, and history of the New World. Here, the author recounts the history of Columbus's discoveries on the new continent; the first interactions with the region's natives; and the different animals, flowers, and trees endemic to the area. He also makes mention of the riches that Spanish colonizers found there and took back to their country.[87] The passages on the New World's riches, on the region's vegetation and animals, and on Columbus's voyages and interactions with the natives are adapted from four different works: Francisco López de Gómara's two-volume work, *Historia general de las Indias* (1552); Gonzalo Fernández de Oviedo u Valdés's *De la natural hystoria de las Indias* (1535); Pietro Martire d'Anghiera's *De orbe novo decades* (1530); and Agustín de Zárate's *Historia del descubrimiento y conquesta del Peru* (1555).[88] It is clear that the author referred to the Italian translations of these works, since knowledge of Spanish was not very common outside of the Safardim community in the Ottoman Empire.[89] In addition to these European sources, the author made use of other works, evidenced by his frequent quoting of the Arab geographers al-Masudi (d. 956), Zakariya Qazwini (d. 1283), and al-Wardi (d. 1457).

---

86 On the contemporary Spanish sources of *History of the India of the West*, see Thomas Goodrich, "The Search for the Sources of the Sixteenth-Century Tarih-i Hind-i Garbi," *Bulletin of Research in the Humanities* 85 (1982): 269–94. On the connection between *History of the India of the West* and Ottoman interest in the Indian Ocean, see Casale, *The Ottoman Age of Exploration*, 160–63. For a comparative analysis of the work in conjunction with contemporary works from Mexico, see Serge Gruzinski, *What Time Is It There?*, trans. Jean Birrell (Cambridge and Malden: Polity Press, 2010).

87 *Tarih-i Hind-i Garbi*, TKS. Revan Köşkü 1488, 40b–48a.

88 Goodrich, "The Search for the Sources of the Sixteenth-Century Tarih-i Hind-i Garbi," 269–94. Ian Manners suggests that the author was also familiar with Giovanni Battista Ramusio's *Delle Navigationi et Viaggi* (1550–59), which includes a reproduction of two of these accounts. Ian Manners, *European Cartographers and the Ottoman World: Maps from the Collection of O.J. Sopranos* (Chicago: Oriental Institute Museum of the University of Chicago, 2007), 36. On *History of the India of the West* also see, Baki Tezcan, "The 'Frank' in the Ottoman Eye of 1583," 267–96.

89 Ibid.; Faroqhi, *The Ottoman Empire and the World Around It*, 198.

While the author directly quotes these traditional Muslim and contemporary European accounts, he also adds his own comments in several places. For example, while reporting about Columbus's petition to the Spanish king for financial support for his journey, the author says that the king promised to finance Columbus's expedition on condition that he succeeded in conquering the city of Granada.[90] When referring to the Reconquista of 1492, the author lectures his readers about the futility and dangers of conflict among Muslims, as this led to the Muslims' destruction at the hands of Christians:

> Thousands of regrets and sorrows that, due to the wrong decisions, arguments, and strife of Muslim notables, several hundred thousand eminent personages of the Muslims, great religious scholars, and pious and righteous ones were ruined in the hands of the despicable non-believers. Even now many thousands of Muslims and pious ones suffer in the hands of wickedness, and voted for the wretchedness of the humility of being a subject.[91]

In the same passage, he goes so far as to urge a leader among the Muslims to "raise a banner of endeavor and resolution, that in the aforementioned marches He may spread the seeds of jihad and war."[92] The author seems to be entreating the Ottoman sultan to unite the Muslims under his banner and save them from doom. A similar passage about the conquest of Mexico implores the sultan to take action. After describing how the Spanish invaded Mexico, sent the region's riches (including gold, pearls, and emeralds) to Spain, and distributed the land among Spanish commanders, the author has this to say to the Ottoman sultan:

> It is requested of His Glorious Majestic Excellency that in the future the bloodthirsty sword of the people of Islam reach that advantageous land, and that its regions and districts be filled with the lights of the religious ceremonies of Islam, and that the possessions and goods that have been mentioned and the other treasures of the unbelievers marked by disgrace be divided, with the permission of the Lord God, among the masters of the Holy War and the nation full of dividing force.[93]

History of the India of the West was more than just an attempt to assemble recent knowledge about the New World for the Ottoman audience. While the author compiled his work by combining the traditional Islamic accounts of "marvels" and "wonders" with contemporary geographic accounts circulating in Europe, he did not refrain from adding his own comments, advising the sultan to protect Muslims the world over and to conquer the rich lands of the New World. We know very little about the author or his patron. It is clear that the work was prepared

---

[90] *Tarih-i Hind-i Garbi*, TKS. Revan Köşkü 1488, 41b.

[91] Ibid., 41b. Goodrich, *The Ottoman Turks and the New World*, 151.

[92] Ibid.

[93] *Tarih-i Hind-i Garbi*, TKS. Revan Köşkü 1488, 85b. Goodrich, *The Ottoman Turks and the New World*, 253.

under the patronage of an Ottoman court elite in an effort to change the Ottoman imperial policy of the time and to adopt a more active stand in international politics. The author here reminds his readers of the sultan's role as protector of all Muslims. These passages, which urge the sultan to fulfill his obligations, should be read within the context of the 1580s, when the work was prepared. This was a period of political and economic crisis. Written during Murad III's heated conflict with the Safavids, the work clearly tried to encourage the sultan to be more assertive in defending his claim to be the leader of all Muslims. In the end however, the sixteenth-century Ottoman court did not take the anonymous author's recommendations seriously and did not incorporate the work or the New World into its imperial policies.[94]

History of the India of the West survives today in 19 manuscript copies and in two printed editions.[95] Among the extant manuscripts, only two are from the sixteenth century, while eight are from the seventeenth century. Most of the copies include illustrations, and at least one or two include world maps.[96] The illustrations depict the plants and creatures described in the text. It was common to include pictures of plants, animals, and strange creatures in Muslim and European accounts of the New World. So in this respect, the author of *History of the India of the West* followed tradition. The work—in its manuscript and printed editions—seems to have been the most successful of such works in transmitting useful political and geographical knowledge about the New World to intellectual and political circles in the Ottoman Empire in the next centuries.

These geographical works coincide with a period when the Ottoman claims for universality were seriously challenged by the Spanish in the Mediterranean and the Portuguese in the Indian Ocean. In this particular period, the Ottoman court welcomed geographical and historical accounts on distant parts of the world, in particular on India, the Indian Ocean, and the New World. In their accounts, Ottoman geographers reported on the recent political developments in these regions. They also discussed the latest geographical knowledge related to these areas as well as their history, inhabitants, customs, and natural environments. As in the case of India, the authors took great pains to report on the Muslim communities in the region. They sometimes directly encouraged the incumbent sultan to launch military campaigns in these distant regions.

Despite the efforts by these individual geographers and their brave patrons to turn the Ottoman state's political and ideological focus to these regions, the leading imperial agenda at the Ottoman court did not envisage navigating in these new waters and thus projected these regions onto the peripheries of the

94 See Tezcan, "The 'Frank' in the Ottoman Eye of 1583," 280–85.

95 For the location and dates of the copies, see Goodrich, *The Ottoman Turks and the New World*, 21–9.

96 The earliest copy, from 1583–84, includes two maps. This is the edition published by İbrahim Müteferrika in 1730. The edition includes 13 woodcut pictures and two world maps, one depicting the whole world and the other showing only the Western Hemisphere. Goodrich, *The Ottoman Turks and the New World*, 28.

Ottoman geographical consciousness and the imperial enterprise. In other words, these regions, their history and geography were important for the Ottoman court to learn about. However, geographical knowledge about these regions and the European accomplishments were only of secondary importance to the Ottoman accomplishments in the Mediterranean. As opposed to the Mediterranean—where the Ottoman state chose to exert all its resources to achieve naval domination, economic control, and imperial sovereignty—India, the Indian Ocean, and the New World stood on the periphery of the Ottoman world throughout the sixteenth century, where territorial and economic expansion were not feasible.

In these distant regions, where the Portuguese and Spanish were keen to expand economically and territorially, the Ottoman presence never had an expansionist dimension. In the case of India and the Indian Ocean, the Ottomans were aware of the region's importance in the spice trade. The Ottoman fleet was active in the area to keep the Red Sea, the Persian Gulf, and their trade networks safe from Portuguese attacks. Unlike the Portuguese, however, the Ottomans did not seek territorial or commercial expansion in the region. Their aim was two-pronged: to defend what they already had, and to put military pressure on the Safavids by controlling their southern and eastern borders.

In the case of the New World, the Ottomans' interest did not extend beyond that of keeping up with current geographical knowledge pertaining to the region. The Ottomans were never as interested in the economic resources of the New World as the Spanish or British monarchies. Contrary to widespread belief, however, the Ottoman court was open to learning about the New World. The number of sixteenth-century Ottoman geographical and historical works on India and the New World is negligible, especially in comparison to works on the Mediterranean or Constantinople. Nevertheless, they reflect Ottoman geographers' attempts to meet the demands and desires of the Ottoman court. These geographers sought the patronage of the court by presenting the latest knowledge about these regions through the lens of Ottoman imperial claims. In doing so, they defined more clearly the borders and "peripheries" of the Ottoman world.

Epilogue
Ottoman Geographical Knowledge in the Long Eighteenth Century

In 1680, Luigi Ferdinando Marsigli, a nobleman, soldier, and cartographer from Bologna, visited Constantinople as part of the Venetian envoy's party. While in Constantinople, accompanied by his Jewish interpreters or the dragomans of the Venetian embassy, Marsigli gathered notes on a variety of topics, including the Ottoman office holders, military, taxation, and medical virtues of coffee. He also became friends with physicians, historians, and geographers who had close contacts with the imperial court, and tried to collect information on Ottoman maps and charts. Marsigli's role and importance for European military mapping and early thematic mapping are well established.[1] His short stay in Constantinople and his contacts with the leading figures of Ottoman intellectual circles also offer insights into Ottoman geographical knowledge in the long eighteenth century. This is what this chapter is concerned with.

By the late sixteenth century, it became clear that the Ottoman imperial enterprise had failed to fulfill its aspirations. As the age of Ottoman territorial expansion drew to a close, the court continued to build on the administrative and bureaucratic structures that were established and consolidated in the previous century. When Mehmed III (r. 1595–1603) ascended to the throne in 1595, he inherited conflicts on two fronts, financial problems, and a rebellion in Anatolia. In the East, Shah Abbas of the Safavids entered Tabriz in 1603 and threatened Ottoman territories in Azerbaijan and Georgia. After the shah invaded Shirvan in 1606, he recovered all the territories that the Safavids had lost in the war of 1578–90.[2]

In Southeastern Europe, meanwhile, the Ottomans had been at war with the Habsburgs since 1593. The new sultan personally led his army to Hungary in 1596, reviving a tradition that had ended during a military campaign in 1566 with

1 On Marsigli's life, travels, and contributions to European cartography, see Anita McConnell, "L. F. Marsgili's Voyage to London and Holland, 1721–1722," *Notes and Records of the Royal Society of London* 41, no. 1 (1986): 39–76; John Stoye, *Marsigli's Europe 1680–1730: The Life and Times of Luigi Ferdinando Marsigli, Soldier and Virtuoso* (New Haven and London: Yale University Press, 1994); M. Cavazza, "The Institute of Science of Bologna and the Royal Society in the Eighteenth Century," *Notes and Records of the Royal Society of London* 56, no. 1 (2002): 3–25; Zsolt Török, "Luigi Ferdinando Marsigli (1658–1730) and Early Thematic Mapping in the History of Cartography," in *Térkép – tudomány / Map – Science*, ed. L. Zentai, J. Györrfy, and Zs. Török (Budapest: ELTE Térképtudományi és Geoinformatikai Tanszék, 2006), 403–13.

2 Imber, *The Ottoman Empire*, 72.

the death of Süleyman I. The Long War, as it was called, raged for 13 years, finally ending during the reign of Ahmed I (r. 1603–17) with the Treaty of Zsitvatorok. Through this treaty the Ottomans gained a few fortresses, but lost territory in Hungary.[3] They also now had to refer to the Habsburg sovereign as "Emperor," thus recognizing his equal status with the sultan.[4] Despite the compromises made in this treaty, it brought a long peace to the western frontier until the 1660s.

The Long War with the Habsburgs coincided with the Jalali rebellions in Anatolia.[5] By 1599, the leader of the rebellions, Abdülhalim Karayazıcı—a former commander of a *bölük* (squadron) of the Malatya *sancakbeyi* (Sanjak governor)—controlled most of Anatolia. To the Ottoman court's annoyance, he claimed that he was descended from the line of former shahs and that the Prophet had granted him the right to rule. The Ottoman court also received reports that Karayazıcı had issued an imperial seal and learned that he had appointed judges and a grand vizier for himself, and had started recruiting soldiers.[6] Even after his death in 1602, the Ottoman state had a difficult time recovering its authority in Anatolia, and didn't bring a decisive end to the rebellions until 1610.[7] Soon after the end of the rebellion, Sultan Ahmed I ordered the construction of the Blue Mosque (the Sultan Ahmed Mosque) in Constantinople to commemorate the Ottoman state's victory over the rebels.

As for the functioning of the government machinery, as financial problems became chronic, the financial offices of the bureaucracy gained more importance. Careers in finance began to offer more mobility and advancement than careers in the scribal offices. As the *damad* (son-in-law) vizier households and factionalism became important sources for individual promotion in the bureaucracy, and as territorial expansion came to a halt, the offices of the bureaucracy became more and more crowded. The new system of appointment by contract shortened one's tenure in office, and the appointment process in general became rather disorganized.[8]

Together with military, social, economic, and administrative adversities, deficiencies in the government machinery, in the army, and in social life were observed and recorded by contemporaries. Many took notice of the decline in the

---

3 Ibid., 68–72.

4 Before Zsitvatorok, the Ottoman court designated the Habsburg ruler the "King of Vienna" and thus assigned a lower status to the Habsburg sovereign than to the Ottoman sultan.

5 On the Jalali rebellions, see Mustafa Akdağ, *Celali İsyanları, 1550–1603* (Ankara: Ankara Üniversitesi Dil ve Tarih-Coğrafya Fakültesi Yayınları, 1963); Michael Cook, *Population Pressure in Rural Anatolia, 1450–1600* (London: Oxford University Press, 1972); Halil İnalcık, "Military and Fiscal Transformation in the Ottoman Empire," *Archivum Ottomanicum* 6 (1980): 283–337.

6 Akdağ, *Celali İsyanları, 1550–1603* (Ankara: Ankara Üniversitesi Dil ve Tarih-Coğrafya Fakültesi Yayınları, 1963), 199, 246–9; Cook, *Population Pressure in Rural Anatolia*, 36; Baki Tezcan, *The Second Ottoman Empire: Political and Social Transformation in the Early Modern World* (Cambridge: Cambridge University Press, 2010), 65.

7 Imber, *The Ottoman Empire*, 73–6.

8 Fleischer, *Bureaucrat and Intellectual*, 214–31.

functioning of the state and its administration, and proposed remedies for these problems.[9] In doing so, they created the notion of a bygone "Golden Age" that had existed during the reigns of Mehmed II and Süleyman I. In their works, they tried to encourage (and sometimes force) the incumbent sultan to revive those glorious days.[10] Until very recently, historical literature on the Ottoman Empire continued to adopt this periodization.[11] A new wave of Ottoman studies, however, has demonstrated not only that there was no golden age in the sixteenth century, but that the eighteenth century was by no means a period of decline.[12] According to this literature, starting in the seventeenth century, the Ottoman Empire underwent a transformation rather than decline in the state administration and political economy.[13] For instance, we now know that the Ottoman economy eventually adapted to the new developments in the world economy, in particular to the opening of new trade networks in the Atlantic Ocean. Moreover, as the office of the sultan and his household lost its primacy in the administration of the state, grand viziers as well as military and religious officials and their households all claimed a share in governing the state. As a result, intellectual production increased in this period under the patronage of an increasing number of elite households. Close analysis of the numbers and the changing nature of the geographic works from this period will further challenge the traditional accounts of the Ottoman Empire and offer valuable insights about the transformation of Ottoman political and intellectual culture during the long eighteenth century.

---

9 For a historical analysis of the contemporary literature on the decline of the Ottoman Empire, see Bernard Lewis, "Ottoman Observers of Ottoman Decline," *Islamic Studies* 1 (1962): 71–87; Cemal Kafadar, "The Question of Ottoman Decline," *Harvard Middle Eastern and Islamic Review* 4, no. 1–2 (1997–98): 30–76; Halil İnalcık, "Military and Fiscal Transformation in the Ottoman Empire," 283–337; Douglas Howard, "Ottoman Historiography and the Literature of 'Decline' of the Sixteenth and Seventeenth Centuries," *Journal of Asian History* 22, no. 1 (1988): 52–77.

10 Cornell Fleischer, "From Şehzade Korkud to Mustafa Ali: Cultural Origins of the Ottoman Nasihatname," in *Third Congress of the Social and Economic History of Turkey, Princeton University, 24–26 August 1983*, ed. Heath Lowry and Ralph S. Hattox (Istanbul: ISIS Press, 1990), 67–77.

11 Bernard Lewis, *The Emergence of Modern Turkey* (Oxford: Oxford University Press, 1968), 21–39; Norman Itzkowitz, *Ottoman Empire and Islamic Tradition* (Chicago: University of Chicago Press, 1972), 67, 73; Halil İnalcık, *The Ottoman Empire: The Classical Age 1300–1600* (New Rochelle: Orpheus Publishing, 1973), 41–52.

12 On the advantages and shortcomings of this new literature in Ottoman studies, see Alan Mikhail and Christine M. Philliou, "The Ottoman Empire and the Imperial Turn," *Society for the Comparative Study of Society and History* 54, no. 4 (2012): 721–45.

13 Recent works by Baki Tezcan and Sam White offer different perspectives on this period. While Baki Tezcan sees the period as that of the Second Empire (rather than as a period of decline), Sam White maintains that the seventeenth century was a period of transformation in the Ottoman Empire and that environmental changes played a major role in the transformation of the political economy of the empire. Tezcan, *The Second Ottoman Empire*; Sam White, *The Climate of Rebellion in the Early Modern Ottoman Empire* (Cambridge: Cambridge University Press, 2011).

After the end of the Long War and the Jalali rebellions in particular, the Ottoman court continued to pursue the imperial tradition of offering patronage to authors of geographical works. From the second half of the seventeenth century on, the number of textual expressions of geography steadily increased in the Ottoman Empire. In the sixteenth century, around 40 geographical accounts were circulating in the empire. In the seventeenth and eighteenth centuries combined, this number reached 100. The majority of these works were produced from the second half of the seventeenth century on.[14] Along with this increase in the number of geographical works came variation in the content of these works. In the eighteenth century, for instance, in addition to travel accounts, works on mathematical geography, and *portolan* charts, translation of European atlases, original works on world geography, cosmographies, as well as ambassadorial accounts, circulated more frequently in Constantinople. Among the 40 works from the sixteenth century are a considerable number of histories and chronicles, which bore geographic descriptions of the Ottoman domains and territories beyond. This stands in direct contrast with a considerable number of works from the eighteenth century that are in fact titled "geography."[15] As the numbers suggest, the members of the imperial court in Constantinople eagerly consumed histories, travel accounts, and translations of contemporary European accounts of geography in the long eighteenth century. At the same time, they threw their support behind the production of original geographical and cartographical works at unprecedented levels. Katip Çelebi (d. 1657), who wrote the first Ottoman work on world geography, *Cihannüma*, and Ebu Bekir ibn Behram ed-Dimashki (d. 1691), who translated Dutch cartographer Joan Bleu's *Atlas Maior* into Ottoman Turkish, were pioneer geographers in this period. These works informed the court of the world's geography, while also helping the ruling elites to reevaluate the role of the Ottoman Empire on the world stage during a period that is typically regarded as the beginning of Ottoman decline.

What accounts for the increase in the number of geographical works produced at this time? Military events like the 25-year conflict over Crete might explain this increase, especially that of cartographical works. The maps still held symbolic and practical value during the battles. Between 1645 and 1669, Venice and the Ottoman Empire engaged in a protracted war over Crete.[16] Although the war resulted in victory for the Ottomans and marked the end of the Latin presence in the eastern Mediterranean, it exhausted the Ottoman economy and the patience of the public. For Venetians, the long war offered an opportunity to keep a fleet stationed in the eastern Mediterranean for almost 25 years. During the war, both sides could keep an eye on trade and politics in the region, on pirates, and on British and Dutch

---

[14] İhsanoğlu, *Osmanlı Coğrafya Literatürü Tarihi*, vol. 1, 74.

[15] Ibid., 73–6.

[16] On the importance of Crete for the early modern Mediterranean basin, and on cultural and economic exchanges on the island both before and after the Ottoman conquest, see Molly Greene, *A Shared World: Christians and Muslims in the Early Modern Mediterranean* (Princeton: Princeton University Press, 2000).

merchant ships, which began to proliferate in the Mediterranean in this period. The encroachment of Atlantic powers on the territory once controlled by the Ottomans also spurred the production of cartographical works in the Ottoman Empire.

The most remarkable work of cartography to come out of this period is an anonymous *portolan* chart depicting the Mediterranean (Plate 4). We know little about this map and can therefore only speculate that it was prepared as a blank map in a workshop in one of the mapping centers in Europe, and that the place names were later added by someone in the Ottoman Empire.[17] The five full and five half wind roses colored in blue, black, and gold hint at a wealthy Ottoman, possibly an elite member of the Ottoman court, as the map's patron.

At first glance, the map does not look any different from the *portolan* charts prepared for the Ottoman court a century earlier. However, when analyzed closely, it becomes clear that this map actually is a turning point in the history of Ottoman geographical knowledge. The map differs remarkably from the charts of the sixteenth century in several areas: first, it offers a more detailed depiction of the rivers and the inland topography. The mapmaker probably knew the topography of the whole region well or at least had in his possession maps or charts of those regions. Secondly, the map marks the major cities around the Mediterranean more comprehensively. In the previous century, only the cities along the coast were labeled with a name or a flag. In this one, even the cities which are not directly on the coast such as Cairo, Aleppo, Sivas, Konya, Sofia, Belgrade, and Vienna are clearly identified with their names in Ottoman-Turkish spelling. Last but not least, the map clearly labels the three major provinces of the Ottoman domains: the province of Anatolia (*vilayet-i Anadolu*), of the Balkans (*vilayet-i Rumili*), and of Africa (*vilayet-i Afrika*). This stands in direct contrast to the maps and charts circulated in the Ottoman Empire in the sixteenth century, none of which mark the administrative division of the imperial domains. The publication date of this anonymous *portolan* map falls right in the middle of the conflict over Crete, suggesting that the cartographer possibly thought that it was a good time to prepare a chart of the Mediterranean. Although it would be wrong to consider the labels for the provinces as border markers, they did probably provide a sense of the extent of the Ottoman domains to an Ottoman court elite.[18]

In 1653, around the same time that this anonymous chart was prepared, Katip Çelebi translated Gerardus Mercator's *Atlas Minor* with the help of Mehmed İlhas,

---

[17] Today, the map is located in the manuscript department of the Bayerische Staatsbibliothek: Bayerische Staatsbibliothek (Handschriften Abteilung), Cod. Turc. 431. See Pinto, "Ways of Seeing."

[18] On the Ottoman borders in Europe especially after the treaty of Karlowitz, see Rifa'at A. Abou-El-Haj, "The Formal Closure of the Ottoman Frontier in Europe, 1699–1703," *Journal of the American Oriental Society* 89, no. 3 (1969): 467–75; Colin Heywood, "The Frontier in Ottoman History: Old Ideas and New Myths," in *Frontiers in Question: Eurasian Borderlands, 700–1700*, ed. Daniel Power and Naomi Standen (New York: St. Martin's Press, Inc., 1999), 228–51.

a former French priest and a convert to Islam.[19] The Topkapı Palace library holds three copies of this translation.[20] Katip Çelebi was a historian, biographer, and geographer. He received a traditional *madrasa* education and joined the chancery as an apprentice in 1622. In 1635, he gave up his career as a scribe and studied numerous topics, including religious sciences, law, mathematics, astronomy, and geography—in particular, cartography—under the patronage of the Şeyhülislam of the period. Quickly, he became a distinguished member of Constantinople's intellectual circles. Western scholars like Antoine Galland and Ferdinando Marsigli are known to have had contacts with these circles. Although neither Marsigli nor Katip Çelebi acknowledges the other, they probably met. Katip Çelebi was one of the most scholarly Ottoman intellectuals. He left 18 works concerning a variety of topics including encyclopedic projects, translations, occasional treatises, and didactic or entertaining compilations.[21]

He prepared his translation of Mercator's atlas between 1653 and 1655, and was probably inspired by the imperial campaign over Crete in preparing this translation.[22] In the first instance, the work provides a more comprehensive and more detailed presentation of the world for its audience than the atlases of the previous century. Katip Çelebi's translation is 429 folios long and it includes 148 maps.[23] It opens with a world map followed by generic maps of each continent, such as Africa, and then more detailed charts of each state in Europe, Asia, Africa, and the New World. While some maps like that of the British Isles and Spain are left empty, others like Japan and South America are very detailed, with place names and topographical indicators. Katip Çelebi probably left the empty maps aside in hopes of filling them out later. Even though Katip Çelebi prepared this translation for his own purposes, his translation was copied several times in the eighteenth and nineteenth centuries. There are 14 copies of this work now extant, five of which are from the eighteenth and nine from the nineteenth century.[24]

---

19 Faroqhi, *The Ottoman Empire and the World Around It*, 199–200; Karamustafa, "Maps and Plans," 218. Katip Çelebi also compiled and translated numerous works on the history and geography of the Ottoman Empire, Europe, and the universe. On Katip Çelebi and his works, see Gottfried Hagen, *Ein osmanischer Geograph bei der Arbeit Entstehung und Gedankenwelt von Katib Çelebis Cihannüma* (Berlin: Klaus Schwarz Verlag, 2003). On Katip Çelebi's presentation of the Pacific and East Asia for his Muslim audience, see John J. Curry, "An Ottoman Geographer Engages the Early Modern World: Katip Çelebi's Vision of East Asia and the Pacific in the Cihânnümâ." *Journal of Ottoman Studies* 40 (2012): 221–57.

20 Goodrich, "Old Maps in the Library of Topkapi Palace in Istanbul," *Imago Mundi* 45 (1993), 125–32. The autograph copy, based on the 1621 Arnheim edition, is kept in the Nuruosmaniye Library in Istanbul. Three other copies are located in the Süleymaniye Library.

21 Gottfried Hagen, "Historians of the Ottoman Empire: Katib Çelebi." Originally published on ottomanhistorians.com, 2007. Last accessed via https://www.academia.edu/3488778/Historians_of_the_Ottoman_Empire_Katib_Celebi#.

22 Ibid.

23 There were 152 maps embedded in the1621 issue of the atlas.

24 İhsanoğlu, *Osmanlı Coğrafya Literatürü Tarihi*, vol. 1, 90–92.

Katip Çelebi's translation differs in content and detail from the charts contained in the three *portolan* atlases prepared for the Ottoman court in the sixteenth century. The atlases from the sixteenth century almost exclusively focused on the Mediterranean and mostly omitted the farther regions like the Indian Ocean and the New World. Their primary aim was to aesthetically depict the important regions of the world for the Ottoman imperial court. Katip Çelebi's translation, on the other hand, represents every map depicted in the original atlas, including the one that charts the religious division of the world according to Christian, pagan, and Muslim populations. The maps in Katip Çelebi's translation have no aesthetic and scientific quality. They bear no signs of having been constructed with mathematical instruments, including those used to create a projectional grid, as in the maps of Europe, Africa, Asia, the Americas, and the Northern and southern poles in the original atlas. The captions for place names on the maps employ the Latin spellings. Katip Çelebi drew the maps probably with Mehmed İlhas.

Katip Çelebi started translating this atlas for himself while researching ways to improve his first encyclopedic work on world geography, *Cihannüma*. In this work, Katip Çelebi initially intended to include recent information on Europe and the New World. However, the geographer gave up on this enterprise because he was dissatisfied with the available sources on Europe and the New World. After he acquired Hondius's edition of Mercator's atlas, he started working on the second version of *Cihannüma*. Among his sources for *Cihannüma* were contemporary European geographical accounts such as *Theatrum Orbis Terrarum* by Ortelius, *Introductio in totam geographiam* by Philippus Cluverius, and a commentary on Aristotle's *Meteorologica* by the Jesuit Collegium Conimbricense, as well as various Ottoman and Islamic geographers and historians such as Abu al-Fida, Aşık Mehmed, Piri Reis, Hoca Sa'deddin, Hamdallah Mustaufi, and Amin Ahmad Razi.[25]

Beyond its political context, Katip Çelebi's *Cihannüma* epitomizes how the Ottoman geographers were active participants in contemporary discussions regarding the usefulness of geographical knowledge and the science of geography. In the introduction to the second version of *Cihannüma*, Katip Çelebi first describes what geography is. He states, "Geography is different from cosmography [...] the science of geography [*ilm-i cografya/fenn-i cografya*] is part of the science of astronomy along with Hydrography, Horography, and Topography."[26] Drawing from the works of previous geographers, Katip Çelebi further divides the science of geography into two sections: mathematical geography and historical geography (*cografya-yi historika*).[27] In this section, Katip Çelebi also argues that the study of geography provides not only a better understanding of the world but also a

---

[25] On *Cihannüma* and its textual analysis, see Franz Taeschner, "Zur Geschichte des Djihannuma," *Mitteilungen des Seminars für Orientalische Sprachen* 2, no. 29 (1926): 99–110; Gottfried Hagen, *Ein osmanischer Geograph bei der Arbeit Entstehung und Gedankenwelt von Katib Çelebis Cihannüma*. For a detailed bibliography on Katip Çelebi and his works see also Hagen, "Katip Çelebi."

[26] Katip Çelebi, *Kitab-ı Cihannüma li Katip Çelebi* (Ankara: TTK, 2009), vol. 1, 15.

[27] Ibid.

strategic and political advantage during military conflicts: "Geography is one of the sciences that are quite beneficial and useful in the civil and social lives of men, and mastering this science is more important than all things else for ministers and senior officials [...] in case of any dispute on the borders of states this science would help resolve a difficulty."[28] Katip Çelebi's compelling argument in favor of geographical knowledge here differs from how sixteenth-century Ottoman intellectuals understood geography. For instance, while most of the geographical works in the sixteenth century were called *history* (*tarih*) and gave no clear definition of geography, already in the second half of the seventeenth century, starting with Katip Çelebi, Ottoman geographers offer a more organized and explicit definition of the science of geography for their readers and emphasize the usefulness of this science in resolving political and military conflicts.

In *Cihannüma*, after the introduction, Katip Çelebi first presents an overview of the seas and the continents, and then continues with descriptions of the countries from east to west. Each description is derived from Mercator's atlas. Each chapter is organized so as to be accompanied by a map. However, there are only 27 maps and 13 plates embedded in the work. Some of these maps were directly copied from the *Atlas Minor* and some were drawn by Katip Çelebi. Although the geographer did not dedicate his work to any Ottoman court member, his other treatises suggest that he had great sympathy for political and religious figures who attempted political reform.[29]

The second version of *Cihannüma* remained unfinished. In addition to the 10 manuscript copies from the seventeenth and eighteenth centuries and 23 from the nineteenth century, *Cihannüma* was also among the first works published by the first Arabic-script printing press in the Ottoman Empire. According to the latest research, *Cihannüma* enjoyed a moderate commercial success.[30] Of the 500 copies of *Cihannüma* printed, 251 were sold.[31] The numbers indicate that the Ottoman court was not the only audience for these works. Katip Çelebi was clearly aiming at educating the Ottoman court and helping to formulate its imperial policies. But his work definitely found a larger audience both within and outside the Ottoman Empire in the following centuries.[32]

The autograph copy of *Cihannüma* from the seventeenth century bears marginal comments by another leading geographer of the time, Ebu Bekir ibn Behram ed-Dimashki (d. 1691). Dimashki was a *madrasa* scholar who earned the patronage of the Grand Vizier Köprülü Fazıl Ahmed Pasha. In 1668, the Dutch envoy in Istanbul, Justinus Colier, presented a copy of the Latin edition of Joan

---

28 Ibid., 16–17.

29 Hagen, "Katip Çelebi."

30 Sabev, "The First Ottoman Turkish Printing Enterprise: Success or Failure," in *Ottoman Tulips, Ottoman Coffee: Leisure and Lifestyle in the Eighteenth Century*, ed. Dana Sajdi (London: I. B. Tauris, 2007), 71.

31 The best seller was the Vankulu Arabic-Turkish dictionary. Of the 500 copies, only one was left unsold. Ibid.

32 Hagen, "Katib Çelebi." See also Hagen, *Ein osmanischer Geograph bei der Arbeit Entstehung und Gedankenwelt von Katib Çelebis Cihannüma*, 415–26.

Bleu's *Atlas Maior* (1662) as a gift to the Ottoman sultan. In 1675, the grand vizier commissioned Dimashki to supervise the translation of this voluminous work, which in the original contained 600 maps. The geographer completed the translation, titled the *Nusretü'l-İslam ve's-Surur fi Takrir-i Atlas Mayur* (*The Triumph of Islam and Joy in the Writing of Atlas Maior*) in 1685. This nine-volume translation, an abridged version containing only 243 maps, was a collective enterprise of a team of translators and cartographers. Ten manuscript copies of this work are housed in different libraries in Istanbul.[33] In his translation, Dimashki amended the original text in accordance with the Islamic geographical works, especially in the chapters where he depicts the Ottoman domains.[34]

Dimashki's translation opens with an introduction where the author makes a distinction between astronomy and geography. He further states that those intellectuals who are familiar with one of the two are also conversant with the other. In this section, Dimashki emphasizes the importance of geographical knowledge. In a similar fashion to Katip Çelebi, Dimashki maintains that the Europeans had gained superiority over the Muslim world thanks to their access to better geographical knowledge. Dimashki, here, introduces Ottoman readers to the full description of Copernicus's heliocentric system.[35] Ebu Bekir Dimashki and Katip Çelebi's analysis of the usefulness of geographical knowledge in the military affairs of states was in fact in tune with the contemporary intellectual discussions taking place in Europe. In the eighteenth century, "geography as a subject and geographical knowledge as a set of intellectual practices concerned with knowing the world" were two separate notions. As "geography" became a serious topic of intellectual discussion in the eighteenth century, contemporary intellectuals and rulers continued to use geographical knowledge to articulate their interests. In addition to their practical value in the battlefields, maps were now important tools to categorize the newly discovered lands and to classify different societies, their cultures, and histories.[36] The eighteenth

33 İhsanoğlu, *Osmanlı Coğrafya Literatürü Tarihi*, vol. 1, 113.

34 Franz Taeschner, "Djughrafiya: The Ottoman Geographers," *EI²*; George Vlahakis, *Imperialism and Science: Social Impact and Interaction* (Santa Barbara, CA: ABC-CLIO, 2006), 81–2.

35 Ekmeleddin İhsanoğlu, "Modern Islam," in *Science and Religion Around the World*, ed. John Hedley Brooke and Ronald L. Numbers (Oxford: Oxford University Press, 2012), 154–5; Behlül Özkan, *From the Abode of Islam to the Turkish Vatan: The Making of a National Homeland in Turkey* (New Haven and London: Yale University Press, 2012), 23–4.

36 For a discussion on "geography" and "geographical knowledge" in the eighteenth century, see Charles Withers, *Placing the Enlightenment: Thinking Geographically about the Age of Reason* (Chicago: University of Chicago Press, 2007); Charles Withers and Robert J. Mayhew, "Geography: Space, Place and Intellectual History in the Eighteenth Century," *Journal for Eighteenth-Century Studies* 34, no. 4 (2011): 445–52. On the importance of the new geographical knowledge for the emergence of the sciences in the eighteenth century, see Felicity A. Nussbaum, ed., *The Global Eighteenth Century* (Baltimore: Johns Hopkins University Press, 2003).

century Ottoman intellectuals followed these discussions closely. The Ottoman court seemed to take the advice of Katip Çelebi and Dimashki seriously. The Grand Vizier Fazıl Mustafa Pasha used parts of Dimashki's translation during the siege of Vienna in 1683.[37] The Ottoman state also started employing maps and charts in order to resolve border disputes after the Treaty of Karlowitz in 1699.[38]

Ebu Bekir Dimashki was among the acquaintances of Marsigli. Marsigli sought him out and probably paid him for instruction in the geography of the Ottoman domains. It is highly possible that the Ottoman geographer told him of the mistakes he found in *Atlas Maior* with regard to the Ottoman domains. Although Marsigli did not follow this lead, he left valuable descriptions of the boundaries of the empire, its peoples, the Christian churches, harem, and the air quality in Constantinople. Upon his return from Constantinople, Luigi Ferdinando Marsigli offered his services to the Habsburg court. After the treaty of Karlowitz in 1699, he served as the imperial boundary commissioner to draw the international Habsburg-Ottoman-Venetian border. He mapped and marked the border in the field with the help of Johann Christoph Mueller. The two cartographers created a geometrical line: a border in its modern sense.[39]

When Marsigli returned to Europe on the termination of his post in Constantinople, among the notes and charts that he had collected at the Ottoman capital was a navigational guide prepared by a certain Seyyid Nuh.[40] Seyyid Nuh named his work *Kitab-ı Bahri'l-Esved ve'l-Ebyaz* (*Book of the Black Sea and the Mediterranean*).[41] Despite its lack of certain identifying details, the work displays similarities to the *Book of Sea Lore* in terms of context and presentation.[42] These similarities led some historians to believe that the *Book of the Black Sea* was just another copy of the *Book of Sea Lore*.[43] In the *Book of the Black Sea and*

---

37 Özkan, *From the Abode of Islam*, 24.

38 The eighteenth-century Habsburg documents on the border disputes over Wallachia, Moldavia, and Transylvania make frequent reference to these Ottoman charts, which the Ottoman state commissioned in order to prove their claims. See Madalina-Valeria Veres, "Redefining Imperial Borders: Marking the Eastern Border of the Habsburg Monarchy in the Second Half of the Eighteenth Century," in *History of Cartography, Lecture Notes in Geoinformation and Cartography*, ed. E. Liebenberg, P. Collier, and Zs. Török (Heidelberg: Springer-Verlag, 2014). I owe special thanks to Madalina Veres who shared these documents with me.

39 Stoye, *Marsigli's Europe 1680–1730*.

40 For a whole collection of Marsigli's notes and charts, see *Catalogo dei Manoscritti di Luigi Ferdinando Marsili*, ed. Lodovico Frati (Firenze: Leo S. Olschki, 1928).

41 F. Babinger, "Seyyid Nuh and His Turkish Sailing Book," *Imago Mundi* 12 (1955): 182–4. Hans Joachim Kissling, *Der See-Atlas des Sejjid Nuh* (Munich: Rudolf Trofenik, 1966).

42 For instance, the pages of the work carry European watermarks. This, however, does not necessarily verify that the work was produced in Europe.

43 Soucek, "Islamic Charting in the Mediterranean," 277; idem, "The 'Ali Macar Reis Atlas' and the Deniz Kitabı," 26–7. Soucek, in fact, argues that Seyyid Nuh is a fictitious character.

the Mediterranean there are 214 colored *portolan* charts depicting the ports and castles of the Black Sea and the Mediterranean. The only known copy of the work is located in the university library in Bologna in the Marsigli collection.[44]

Two questions about this work need to be answered: who prepared it and why? The name "Seyyid Nuh" suggests that the author was a convert. Seyyid Nuh is most likely the person referred to in contemporary sources as Nuh Efendi (d. 1707). Nuh Efendi or Seyyid Nuh, as he calls himself on the first page of the work, was originally a Venetian citizen. He fell in love with the daughter of an Ottoman merchant while studying medicine at the University of Padua, and converted to Islam in order to marry her. He then settled in Istanbul and became the sultan's physician.[45] Why he produced this work is unclear. Perhaps Marsigli commissioned him to copy Piri Reis's *magnum opus*. Perhaps, Seyyid Nuh was trying to impress the Ottoman court with yet another copy of the *Book of Sea Lore* and to confirm his loyalty to the Ottoman Empire during its war with Venice, his hometown.

These works are only the tip of the iceberg. From the second half of the seventeenth century well into the eighteenth century, the number of geographical works produced and circulated in the Ottoman Empire increased steadily. This increase in number was the result of the production of multiple manuscript copies as well as printed editions. İbrahim Müteferrika, a Transylvanian convert, opened the first Arabic-script printing press in the empire in 1722. Müteferrika printed 24 titles before his death in 1745. The initial print run for each book varied between 500 and 1,000.[46] Among these titles were chronicles, dictionaries, and geographical works, including maps.[47] Müteferrika was indeed keen on printing geographical works. Among the first printed works by Müteferrika were stand-alone maps, *The History of the India of the West*, Katip Çelebi's *Cihannüma*, and *Tuhfetü'l-Kibar fi Esfari'l-Bihar*.[48]

Along with this increase in the number of geographical works came variation in the content of these works. In the eighteenth century, for instance, in addition to the more common works of geography—such as travel accounts and *portolan* charts—atlases of the Ottoman domains and other parts of the world, as well as ambassadorial accounts, circulated more frequently in the empire. This is indicative of two independent but correlated developments. First, it tells us that the Ottoman court was not the only audience for these works. Until the sharp increase in the number of private and public schools in the nineteenth century, only a minority

---

[44] Seyyid Nuh Efendi, *Kitab-i Bahri'l-Esved ve'l-Ebyaz*. Bologna, Marsigli, nr. 3609.

[45] Michael Robert Hickok, *Ottoman Military Administration in Eighteenth-Century Bosnia* (Leiden: Brill, 1997), 17.

[46] Orlin Sabev, who argues that Müteferrika's enterprise was indeed a success, maintains that 500 is the conservative and 1,000 is the liberal estimate for initial print-runs. See Sabev, "The First Ottoman Turkish Printing Enterprise: Success or Failure," 72.

[47] Günay Kut, "Matba'a," *EI*[2], last accessed via Brill Online. University of Pittsburgh. 7 November 2011.

[48] Sabev, "The First Ottoman Turkish Printing Enterprise: Success or Failure," 71.

could read and write in the Ottoman Empire.[49] Therefore, the audience for these works was mainly administrative and military officials, religious functionaries, students, other scholars, as well as non-scholarly educated townsmen (traders and craftsmen) in the long eighteenth century.[50] Nevertheless, it is clear that geography entered the libraries of a wider public in this period. This development marks the beginning of an enlightenment in geographical knowledge in the Ottoman Empire.[51] In the absence of a long-standing print culture, the production of numerous manuscript copies made the diffusion of ideas possible.

Second, the variation in subject matter in the geographical works created at this time suggests that geographical topics that used to be considered part of the same science had become divorced from each other and had grown into separate branches. For instance, cosmography—in other words, the integrated study of earthly and celestial spheres—came to focus exclusively on mathematical cartography and on other complementary empirical practices in the Ottoman Empire. Similar developments in the field of geography were witnessed in the English, French, and Spanish empires during the early modern period.[52]

In 1802, a printing house in Constantinople published the Ottoman Turkish translation of an edition of William Faden's *General Atlas* titled *Cedid Atlas Tercümesi* (*Translation of the New Atlas*). The private secretary to the Ottoman ambassador in London brought the atlas to Constantinople. The intended audience for the atlas was the students and faculty in the recently founded Ottoman Military Engineering School, who were to be trained in contemporary European cartographic practices. Fifty copies of the atlas were printed including a presentation copy for the Ottoman sultan. The translation replicated and engraved all the original maps of Faden's *General Atlas*. The maps detailed the borders of the European nation states and indicated the topographical characteristics of the regions. The historical literature on the development of cartography in the

---

[49] According to the estimates, the literacy rate rose from 2–3 percent in the early nineteenth century to 15 percent at the end of the century. Donald Quataert, *The Ottoman Empire, 1700–1922* (Cambridge: Cambridge University Press, 2005), 169.

[50] Faroqhi, *The Ottoman Empire and the World Around It*, 199. See also Hagen, *Ein osmanischer Geograph bei der Arbeit Entstehung und Gedankenwelt von Katib Çelebis Cihannüma*; Sabev, "The First Ottoman Turkish Printing Enterprise: Success or Failure," 77.

[51] For a discussion on "geography" in enlightenment literature, see Withers and Mayhew, "Geography: Space, Place and Intellectual History in the Eighteenth Century," 445–6.

[52] For a general discussion on the changing nature of cosmography during the early modern period, see Klaus A. Vogel, "Cosmography," in *The Cambridge History of Science, Vol. 3 Early Modern Scicence*, ed. Lorraine Daston and Katherine Park (Cambridge: Cambridge University Press, 2006), 470–71. For a study of French cosmography, see Frank Lestringant, *Mapping the Renaissance World*, trans. David Fausset (Cambridge: Polity, 1994), 129. For the English case, see Lesley B. Cormack, *Charting an Empire* (Chicago: University of Chicago Press, 1997), 37–42. The Spanish case is discussed by Maria M. Portuondo, *Secret Science: Spanish Cosmography and the New World* (Chicago: University of Chicago Press, 2009), 257–91.

Ottoman Empire has thus far regarded the atlas as the turning point in the history of Ottoman geographical knowledge. This atlas, according to this view, was the first time Ottoman cartographers adopted what European mapmakers and map users promoted as a scientific model of collecting, measuring, and presenting geographic information.[53]

This chapter proposes that this process had already started in the second half of the seventeenth century. In this period, Constantinople welcomed intellectuals from the rest of the world who exchanged ideas and knowledge with their Ottoman colleagues, and Ottoman geographers produced works in increasing numbers. In this context, Ottoman geographers not only re-presented contemporary European geographical works but also participated in the current discussions and contributed with their own ideas. In doing so, they informed their audience, including the Ottoman court as well as the larger audience of urban literati, of the usefulness of the science of "geography," and the importance of geographical knowledge. They also advocated the divorce of geography from history, cosmography, and astronomy and offered a narrower description of what the science of geography was. It is clear that Ottoman geography had come a long way from the anonymous *portolan* chart of 1652 to this atlas from 1802. However, the process had already started in the second half of the seventeenth century.

These developments in the seventeenth and eighteenth centuries would not have been possible without the efforts and courage of figures like Piri Reis, Mustafa b. Ali, Seydi Ali Reis, Matrakçı Nasuh, and the many others who strived to turn the Ottoman court's attention to the subject of geography and geography of the world. These sixteenth-century Ottoman geographers ushered in a new era in Ottoman history by emphasizing the practical and symbolic importance of geographical works for the Ottoman imperial enterprise. They did so without formal training or an institutional framework. The efforts of these geographers could have easily been dismissed and their works could have easily disappeared in the vaults of the Topkapı Palace. However, it did not happen that way. These geographers of the sixteenth century valiantly introduced and reintroduced topics of geography and geographical knowledge to the elites of the Ottoman court. By displaying their skills in mapmaking and knowledge about the world's geography, they sought the patronage of those elites. Consequentially, they tried to shape the political and economic policies of the Ottoman court. Their efforts did not go unnoticed. In fact, they became as instrumental to the imperial court as the Ottoman troops in articulating the Ottoman universal claims to the whole world. When Piri Reis presented his world map to Selim I, he was probably not aware that he was initiating a new era in Ottoman intellectual history. His bravery and works paved the way for Katip Çelebi and Ebu Bekir Dimashki, who left behind the most ambitious works of geography, and who ultimately succeeded in convincing both Ottoman society and the Ottoman court of the importance and value of formal engagement with "geography" and "geographical knowledge."

---

[53] Ian Manners, *European Cartographers and the Ottoman World*, 52.

Bibliography

Published and Unpublished Ottoman and European Manuscript Sources

6 Numaralı Mühimme Defteri: 972 / 1564–65, edited by Hacı Osman Yıldırım. Ankara: Başbakanlık Devlet Arşivleri Genel Müdürlüğü Osmanlı Arşivi Daire Başkanlığı, 1995.

'Abdü'l-Latif Çelebi (Latifi). *Evsaf-ı İstanbul*. Istanbul University Library, TY. 3751.

———. *Risale-i Ta'rifname-i İstanbul*. Süleymaniye Library, Esad Efendi 2272.

———. *Evsaf-ı İstanbul*, edited by Nermin Suner (Pekin). Istanbul: Baha Matbaası, 1977.

———. *Tezkire-yi Latifi*. İstanbul: İkdam Matbaası, 1896–97.

Ali Ekber Khitayi. *Khitay-nameh*. Süleymaniye Library, Aşir Efendi 249.

———. *Khitay-nameh: Sharh-i Mushahidat-i Sayyid 'Ali Ekber Khita'i: Mu'asir-i Shah İsma'il Safavid dar Chin*, edited by Iraj Afshar. 2nd ed. Tehran: Markaz-i Asnad-i Farhang-i Asya, 1993.

———. *The Book on China: Khitaynama*, edited by Fuat Sezgin and Eckhardt Neubauerm. Frankfurt: IGAIW, 1994.

Aşık Çelebi. *Meşa'irü'ş-Şu'ara*, edited by G. M. Meredith-Owens. Cambridge: E. J. W. Gibb Memorial, 1971.

Aşık Mehmed. *Menazirü'l-Avalim*, edited by Mahmut Ak. 3 vols. Ankara: TTK, 2007

Bağdadlı İsmail Pasha, *İzah al-Maknun*. Istanbul, 1945–47.

Bostan Çelebi. *Gazavat-ı Sultan Süleyman*. Süleymaniye Library, Ayasofya 3117.

Bursalı Mehmed Tahir. *Osmanlı Mü'ellifleri*. 3 vols. Istanbul: Matba'a-i Amire, 1915–25.

Celalzade Mustafa. *Me'asir-i Selim Hani: Selim-name*. TKS. Revan Köşkü 1274.

———. *Me'asir-i Selim Hani: Selim-name*, edited by Ahmet Uğur and Mustafa Çuhadar. Ankara: Kültür Bakanlığı, 1990.

Davis, William Stearns, ed. *Readings in Ancient History: Illustrative Extracts from the Sources*. 2 vols. Boston: Allyn and Bacon, 1912–13.

Derschwam, Hans. *Tagebuch einer Reise nach Konstantinopel und Kleinasien*, edited by Franz Babinger. Munich and Leipzig, 1923.

Evliya Çelebi, *Evliya Çelebi Seyahatnamesi*, edited and translated by Orhan Şaik Gökyay. 10 vols. Istanbul: Yapı Kredi Yayınları, 1995.

Feridun Ahmed Beg. *Münşeatü's-Selatin*. 2 vols. Istanbul, 1858.

Geographi Latini Minores, edited by Alexander Riese. Heilbronn: Apud Hennineros Fratres, 1878.

Hoca Sa'deddin. *Tacü't-tevarih*, edited by İsmet Parmaksızoğlu. 5 vols. Ankara: Kültür Bakanlığı, 1992.

Idris-i Bidlisi. *Hasht Bihisht*. Süleymaniye Library, Esad Efendi 2198.
İshak Üskübi. *Selimname*. Süleymaniye Library, Hekim Ali Paşa 761.
Katip Çelebi. *Tuhfetü'l-Kibar fi Efsari'l-Bihar*. TKS. Revan Köşkü 1192; Süleymaniye Library, Esad Efendi 2207.
———. *Kitab-ı Cihannüma li Katip Çelebi*. Ankara: TTK, 2009.
Kemal Paşazede (Ibn Kemal). *Tevarih-i Al-i Osman*, edited by Şerafettin Turan. 10 vols. Ankara: TTK, 1954.
Kemal Paşazade. *Tevarih-i Al-i Osman*, edited by Şefaettin Severcan. Ankara: TTK, 1996.
Lokman b. Seyyid Hüseyin. *Şahanşah-name*. 2 vols. Istanbul University Library, F 1404 (vol. 1) and TKS. Bağdad 200 (vol. 2).
Matrakçı Nasuh. *Beyan-ı Menazil-i Sefer-i 'Iraqeyn-i Sultan Süleyman Han*. Istanbul University Library, TY. 5964.
———. *Beyan-ı Menazil-i Sefer-i Irakeyn-i Sultan Süleyman Han*, edited by Hüseyin Yurdaydın. Ankara: Türk Tarih Kurumu, 1976.
———. *Fethname-i Karabuğdan*. TKS. Revan Köşkü 1284/2.
———. *Süleymanname*. TKS. Revan Köşkü 1286.
———. *Tarih-i Feth-i Sikloş ve Estergon ve İstun-Belgrad*. TKS. Hazine 1608.
———. *Tarih-i Sultan Bayezid*. TKS. Revan Köşkü 1272.
Matuz, Joseph, ed. *L'Ouvrage de Seyfi Çelebi: Historien Ottoman du XVIe Siecle*. Paris: Dépositaire, 1968.
Mustafa b. Ali al-Muvakkit. *İ'lamu'l-'İbad fi A'lami'l-Bilad*. Süleymaniye Library, Haci Mahmud 5633.
———. *Tuhfatü'z-Zaman ve Haridatü'l-'Avan*. Süleymaniye Library, Halet Efendi 366.
Mustafa Sa'i Çelebi. *Tezkiretü'l-Bünyan*. Süleymaniye Library, Hacı Mahmud Efendi 4911.
Mustafa Selaniki Efendi, *Tarih-i Selaniki*. Süleymaniye Library, Esad Efendi 2259.
———. *Tarih-i Selaniki*, edited by Mehmet İpşirli. 2 vols. Ankara: TTK, 1999.
Pires, Tomé. *The Suma Oriental of Tome Pires: An Account of the East, from the Red Sea to China, Written in Malacca and India in 1512–1515 and The Book of Francisco Rodrigues: Pilot-Major of the Armada that Discovered Banda and the Moluccas*, translated and edited by Armando Cortesão. 2 vols. New Delhi: Asian Educational Services, 2005.
Ramazanzade Nişancı Mehmed Pasha. *Tarih-i Nişancı Mehmed Paşa*. Istanbul, 1873.
Seydi Ali Reis. *Kitabü'l-Muhit fi 'İlmi'l-Afak ve'l-Abhur*. TKS. Revan Köşkü 1643; Süleymaniye Library, Nuruosmaniye 2948.
———. *Hindelinden İstanbula*, translated by Mustafa Nihat Özön and Hayrullah Örs. Ankara: Köyhocası Matbaası, 1935.
———. *Book of the Indian Ocean: Kitab al-Muhit*, edited by Fuat Sezgin. Frankfurt: IGAIW, 1997.
———. *Mir'atü'l-Memalik*. TKS. Revan Köşkü 1470.
———. *Mir'atü'l-Memalik*. Istanbul: İkdam Matbaası, 1895.

———. *The Travels and Adventures of the Turkish Admiral Sidi Ali Reis in India, Afghanistan, Central Asia, and Persia, During the Years 1553–1556*, translated by A. Vambéry. London: Luzac & Co., 1899.
———. *Mir'at ül-Memalik (Ülkelerin Aynası)*, edited by Necdet Akyıldız. Istanbul: Tercüman, 1975.
———. *Mir'at ül-Memalik: İnceleme, Metin, İndex*, edited by Mehmet Kiremit. Ankara: Türk Tarih Kurumu, 1999.
Seyyid Muradi. *Gazavat-i Hayreddin Paşa*. TKS. Revan Köşkü 1291.
———. *Gazavat-i Hayreddin Paşa: Barbaros Hayreddin Pasa'nin Hatiralari*, translated and edited by Ertugrul Düzdağ. Istanbul: Tercüman Gazaetesi, 1980.
Seyyid Nuh Efendi. *Kitab-i Bahri'l-Esved ve'l-Ebyaz*. Bologna, Marsigli, nr. 3609.
Tarih-i Hindi-i Garbi. TKS. Revan Köşkü 1488.
Tarih-i Hindi-i Garbi veya Hadis-i Nev: History of the West Indies Known as the New Hadith, edited by Tülay Duran, translated and transcribed by Fuat Yavuz, Ayhan Özyurt, Tevfik Temelkuran. Istanbul: Tarihi Araştırmalar Vakfı, Istanbul Araştırma Merkezi, 1999.
Waldseemüller, Martin. *Cosmographiae introductio cum quibus dam geometrieae ac astronomiae principiis ad eam rem necessariis*. St. Die, France, 1507; n.p.: Readex Microprint, 1966.
Yazıcıoğlu Ahmed Bican. *Dürr-i Meknun: Saklı İnciler*, translated by Necdet Sakaoğlu. Istanbul: Tarih Vakfı Yurt Yayınları, 1999.

Published and Unpublished Ottoman and European Maps, *Portolan* Charts, Atlases, and *Isolarii*

Al-Hacc Abu'l-Hasan. *Portolan Chart*, TKS. Hazine 1822
Ali Macar Reis. *Atlas*. TKS. Hazine 644.
Anonymous Portolan Chart. Bayerische Staatsbibliothek (Handschriften Abteilung), Cod. Turc. 431.
Atlas-ı Hümayun (Imperial Atlas). Istanbul Archeology Museum 1621.
Francesco Berlinghieri. *Atlas*. TKS. G. I. 84.
Hacı Ahmed. *Mappamundi*. Bodleian, March, nr. 454.
Menemenli Mehmed Reis. *Portolan Chart*. Ms. Venedig, Museo Correr, Port 22.
Piri Reis. *Mappamundi*. TKS. Revan Köşkü 1633 mük.
———. *Piri Reis Haritasi Hakkında Izahname*, edited by Yusuf Akçura. Ankara: TTK, 1992.
———. *Kitab-ı Bahriye*. 1st ed. TKS. Hüsrev Paşa 272.
———. *Bahrije, Das türkisches Segelhandbuch für das Mittelandische Meer von Jahre 1521*, edited by Paul Kahle. 2 vols. Berlin: Walter de Gruyter, 1926.
———. *Kitab-ı Bahriye*. 2nd ed. Süleymaniye Library, Ayasofya 2612.
———. *Kitab-ı Bahriye*, edited by E. Zekai Ökte, Vahit Çabuk, Tülay Duran, and Robert Bragner. 4 vols. Ankara: Ministry of Culture and Tourism of the Turkish Republic, 1988.
Walters Sea Atlas. Walters Art Museum W 660.

Secondary Sources

Abou-El-Haj, Rifa'at Ali. "The Formal Closure of the Ottoman Frontier in Europe, 1699–1703." *Journal of the American Oriental Society* 89, no. 3 (1969): 467–75.

Abulafia, David. "What Is the Mediterranean?" In *The Mediterranean in History*, edited by David Abulafia, 11–32. London: Thames & Hudson, 2003.

———. "Mediterraneans." In *Rethinking the Mediterranean*, edited by William V. Harris, 64–93. Oxford: Oxford University Press, 2005.

Adas, Michael. *Machines as the Measure of Men: Science, Technology and Ideologies of Western Dominance*. Ithaca: Cornell University Press, 1989.

Adıvar, A. Adnan. *Osmanlı Türklerinde İlim*. 5th ed. Istanbul: Remzi Kitabevi, 1991.

Afetinan, A. *Life and Works of Piri Reis*, translated by Leman Yolaç and Engin Uzmen. Ankara: Türk Tarih Kurumu, 1987.

Ágoston, Gábor. "Information, Ideology, and Limits of Imperial Policy: Ottoman Grand Strategy in the Context of Ottoman-Habsburg Rivalry." In *Early Modern Ottomans: Remapping the Empire*, edited by Virginia Aksan and Daniel Goffman, 75–103. New York: Cambridge University Press, 2007.

Ahmad, Maqbul. "Cartography of al-Sharif al-Idrisi." In *The History of Cartography, Vol. 2, Bk. 1, Cartography in the Traditional Islamic and South Asian Societies*, edited by J. B. Harley and David Woodward, 156–74. Chicago: University of Chicago Press, 1992.

Akdağ, Mustafa. *Celali İsyanları, 1550–1603*. Ankara: Ankara Üniversitesi Dil ve Tarih-Coğrafya Fakültesi Yayınları, 1963.

Alegria, Maria Fernanda, Suzanne Daveau, João Carlos Garcia, and Francesc Relaño. "Portuguese Cartography in the Renaissance." In *The History of Cartography, Vol. 3, Bk. 1, Cartography in the European Renaissance*, edited by J. B. Harley and David Woodward, 975–1068. Chicago: University of Chicago Press, 2007.

Alexander, Paul J. "The Strength of Empire and Capital as Seen Through Byzantine Eyes." *Speculum* 37, no. 3 (1962): 339–57.

Allouche, Adel. *The Origins and Development of the Ottoman-Safavid Conflict (906–962/1500–1555)*. Berlin: K. Schwarz, 1983.

Andrews, Walter. "The Tezkere-i Şu'ara of Latifi as a Source for the Critical Evaluation of Ottoman Poetry." PhD dissertation: University of Michigan, 1970.

Arbel, Benjamin. "Maps of the World for Ottoman Princes? Further Evidence and Questions Concerning 'The Mappamondo of Hajji Ahmed.'" *Imago Mundi* 54 (2002): 19–30.

Arjomand, Said Amir. *The Shadow of God and the Hidden Imam: Religion, Political Order and Societal Change in Shiite Iran from the Beginning to 1890*. Chicago: University of Chicago Press, 1984.

Arnold, Thomas. *The Renaissance at War*. New York: Harper Collins Publishers, 2006.

Astengo, Corradino. "The Renaissance Chart Tradition in the Mediterranean." In *The History of Cartography, Vol. 3, Bk. 1, Cartography in the European Renaissance*, edited by J. B. Harley and David Woodward, 174–237. Chicago: University of Chicago Press, 2007.

Aubin, Jean. "Études Safavides. I. Shah Isma'il et les notables de l'Iraq persan." *JESHO* 2 (1959): 37–81.

———. "La Politique religieuse des Safavids." In *Le Shiisme imamite*, edited by T. Fahd, 235–44. Paris: Presses Univ. de France, 1970.

———. "L'Avenèment des Safavides Reconsidéré (Études Safavides. III)." *Moyen Orient et Ocean Indien* 5 (1988): 1–130.

Babinger, Franz. *Die Geschichtsschreiber der Osmanen und Ihre Werke*. Leipzig: Otto Harrassowitz, 1927.

———. "Piri Muhyi'd-Din Re'is." *EI*[1].

———. "An Italian Map of the Balkans, Presumably Owned by Mehmed II, the Conqueror (1452–53)." *Imago Mundi* 8 (1951): 8–15.

———. "Seyyid Nuh and His Turkish Sailing Book." *Imago Mundi* 12 (1955): 182–4.

———. *Mehmed the Conqueror and His Time*, translated by Ralph Manheim, edited by William Hickman. Princeton: Princeton University Press, 1978.

Bacqué-Grammont, Jean-Louis. *Les Ottomans, les Safavides, et leurs voisins*. Istanbul: Nederlands Historisch-Archaeologisch Instituut te Istanbul, 1986.

Baker, Alan R. H. *Geography and History: Bridging the Divide*. Cambridge: Cambridge University Press, 2003.

Ballon, Hilary, and David Friedman. "Portraying the City in Early Modern Europe." In *The History of Cartography, Vol. 3, Bk. 1, Cartography in the European Renaissance*, edited by J. B. Harley and David Woodward, 680–704. Chicago: University of Chicago Press, 2007.

Barkan, Ömer Lütfi. "The Price Revolution of the Sixteenth Century." *International Journal of Middle East Studies* 6 (1975): 3–28.

Bazan, G. Guevara. "Muslim Immigration to Spanish America." *Muslim World* 56 (1966): 173–87.

———. "Some Notes for a History of the Relations between Latin America, the Arabs and Islam." *Muslim World* 61 (1971): 284–92.

Beldiceanu-Steinherr, Irène, and Jean-Louis Bacqué-Grammont. "A Propos de quelques causes de malaises sociaux en Anatolie Centrale." *Archivum Ottomanicum* 7 (1982): 71–116.

Beller-Hahn, Ildiko. "Ottoman Perception of China." In *Comite International D'etudes Pre-Ottomanes et Ottomanes, VIth Symposium, Cambridge, 1st–4th July 1984: Proceedings*, edited by Jean-Louis Bacque-Grammont and E. J. van Donzel, 55–64. Istanbul: Divit Press, 1987.

Ben-Zaken, Avner. *Cross-Cultural Scientific Exchanges in the Eastern Mediterranean, 1500–1600*. Baltimore: Johns Hopkins University Press, 2010.

Bethencourt, Francisco. "European Expansion and the New Order of Knowledge." In *The Renaissance World*, edited by John Jeffries Martin, 118–40. New York and London: Routledge, 2007.

Birchwood, Matthew, and Matthew Dimmock, eds. *Cultural Encounters between East and West, 1453–1699*. Cambridge: Cambridge Scholars Press, 2005.

Bisaha, Nancy. *Creating East and West: Renaissance Humanism and the Ottoman Turks*. Pittsburgh: University of Pennsylvania Press, 2004.

Bittner, M., and W. Tomaschek. "Die topographischen Capitel des indischen Seespiegels Mohit, mit einer Einleitung sowie mit 30 Tafeln versehen." In *Studies on the Works of Sidi Ali*. Vol. 16 of *Islamic Geography*, edited by Fuat Sezgin et al., 129–254. Frankfurt: IGAIW, 1992.

Blockmans, Wim. *Emperor Charles V, 1500–1558*, translated by Isola van den Hoven-Vardon. London: Arnold Publishers, 2002.

Bostan, İdris. *Osmanlı Bahriye Teşkilatı: XVII. Yüzyılda Tersane-i Amire*. Ankara: Türk Tarih Kurumu, 1992.

———. "Cezayir-i Bahr-ı Sefid Eyaletinin Kuruluşu (1534)." *İÜEF Tarih Dergisi* 38 (2002–03): 61–77.

———. *Osmanlılar ve Deniz: Deniz Politikaları, Teşkilat, Gemiler*. Istanbul: Küre, 2007.

Brafman, David. "Facing West: The Western View of Islam in Nicholas de Nicolay's 'Travels in Turkey.'" *Getty Research Journal* 1 (2009): 153–60.

Braudel, Fernand. *The Mediterranean and the Mediterranean World in the Age of Philip II*, translated by Sian Reynolds. 2 vols. New York: Harper and Row, 1972.

Brauer, Ralph. "Boundaries and Frontiers in Medieval Muslim Geography." *Transactions of the American Philosophical Society* 85, no. 6 (1995): 3–73.

Brentjes, Sonja. "al-Idrisi." In *Medieval Science, Technology, and Medicine: An Encyclopedia*, edited by Thomas Glick, Steven J. Livesey, and Faith Wallis, 261. New York: Routledge, 2005.

Brice, William, and Colin Imber. "Turkish Charts in the 'Portolan' Style." *Geographical Journal* 144 (1978): 528–9.

Brice, William, Colin Imber, and Richard Lorch. *The Aegean Sea-Chart of Mehmed Reis Ibn Menemenli, A.D. 1590/1*. Manchester: University of Manchester Press, 1977.

Brotton, Jerry. *Trading Territories: Mapping the Early Modern World*. Ithaca: Cornell University Press, 1997.

———. "Printing the World." In *Books and Sciences in History*, edited by Marina Frasca-Spada and Nick Jardine, 35–48. Cambridge: Cambridge University Press, 2000.

———. *The Renaissance Bazaar: From the Silk Road to Michelangelo*. Oxford: Oxford University Press, 2002.

Brummett, Palmira. "The Ottoman Adversary: Rhodes and Ottoman Naval Power." *Historical Journal* 36, no. 3 (1993): 517–41.

———. *Ottoman Seapower and Levantine Diplomacy in the Age of Discovery*. Albany: State University of New York Press, 1994.

———. "The Myth of Shah Ismail Safavi: Political Rhetoric and 'Divine' Kingship." In *Medieval Christian Perceptions of Islam*, edited by John Victor Tolan, 331–59. New York: Garland, 1996.

———. "Imagining the Early Modern Ottoman Space, from World History to Piri Reis." In *The Early Modern Ottomans: Remapping the Empire*, edited by Virginia H. Aksan and Daniel Goffman, 15–58. Cambridge: Cambridge University Press, 2007.

Buisseret, David. "Spanish Peninsular Cartography, 1500–1700." In *The History of Cartography, Vol. 3, Bk. 1, Cartography in the European Renaissance*, edited by J. B. Harley and David Woodward, 1069–171. Chicago: University of Chicago Press, 2007.

———, ed. *Monarchs, Ministers, and Maps: The Emergence of Cartography as a Tool of Government in Early Modern Europe*. Chicago: University of Chicago Press, 1992.

Burckhardt, Jacob. *The Civilization of the Renaissance in Italy*. 1860. New York: Modern Library, 2002.

Burke, Peter. "Presenting and Re-presenting Charles V." In *Charles V, 1500–1558, and His Time*, edited by Hugo Soly, 393–476. Antwerp: Mercatorfonds, 1999.

Bush, Barbara. *Imperialism and Postcolonialism*. London: Pearson Education, 2005.

Campbell, Tony. "Portolan Charts from the Late Thirteenth Century to 1500." In *The History of Cartography, Vol. 1, Cartography in Prehistoric, Ancient, and Medieval Europe and the Mediterranean*, edited by J. B. Harley and David Woodward, 371–464. Chicago: University of Chicago Press, 1987.

Canning, Joseph. *A History of Medieval Political Thought 300–1450*. London: Routledge, 1996.

Casale, Giancarlo. *The Ottoman Age of Exploration*. Oxford: Oxford University Press, 2010.

Cassola, Arnold. *The 1565 Ottoman Malta Campaign Register*. Malta: Publishers Enterprises Group (PEG), 1994.

Cavazza, M. "The Institute of Science of Bologna and the Royal Society in the Eighteenth Century." *Notes and Records of the Royal Society of London* 56, no. 1 (2002): 3–25.

Çetin, Nihad M. "Latifi." *EI*[2].

Çıpa, Hakkı Erdem. "The Centrality of the Periphery: The Rise to Power of Selim I, 1487–1512." PhD dissertation: Harvard University, 2007.

Clark, Alfred. "Medieval Arab Navigation on the Indian Ocean: Latitude Determinations." *Journal of the American Oriental Society* 113, no. 3 (1993): 360–73.

Cook, David. *Studies in Muslim Apocalyptic*. Princeton: Darwin Press, 2002.

Cook, Michael. *Population Pressure in Rural Anatolia, 1450–1600*. London: Oxford University Press, 1972.

Cormack, Lesley B. *Charting an Empire*. Chicago: University of Chicago Press, 1997.

Cosgrove, Denis. *Apollo's Eye: A Cartographic Genealogy of the Earth in the Western Imagination*. Baltimore: Johns Hopkins University Press, 2001.

Curry, John J. "An Ottoman Geographer Engages the Early Modern World: Katip Çelebi's Vision of East Asia and the Pacific in the Cihânnümâ." *Journal of Ottoman Studies* 40 (2012): 221–57.

Darling, Linda. "The Renaissance and the Middle East." In *A Companion to the Worlds of the Renaissance*, edited by Guido Ruggiero, 55–87. London: Blackwell Publishers, 2002.

D'Avezac, Marie Armand Pascal. "Note sur une mappemonde turke du XVIe siècle conservée à la Bibliothèque de Saint-Marc à Venise." In *Reprint of Studies on the Ottoman Cartographers Piri Reis (d. 1554) and Haggi Ahmad (d. about 1560)*. Vol. 22 of *Islamic Geography*, edited by Fuat Sezgin et al., 315–99. Frankfurt: IGAIW, 1992.

Diez, H. F. von. *Denkwürdigkeiten von Asien*. 2 vols. Berlin: Nicolai Verlag, 1815.

Dilke, O. A. W. "Cartography in the Byzantine Empire." In the *History of Cartography, Vol. 1, Cartography in Prehistoric, Ancient, and Medieval Europe and the Mediterranean*, edited by J. B. Harley and David Woodward, 258–76. Chicago: University of Chicago Press, 1987.

Doukas, Michael. *The Decline and Fall of Byzantium to the Ottoman Turks*, translated by H. J. Magoulias. Detroit: Wayne State University Press, 1975.

Dressler, Markus. "Inventing Orthodoxy: Competing Claims for Authority and Legitimacy in the Ottoman-Safavid Conflict." In *Legitimizing the Order: The Ottoman Rhetoric of State Power*, edited by Hakan T. Karateke, 151–73. Leiden: Brill, 2005.

Dursteler, Eric. *Venetians in Constantinople: Nation, Identity, and Coexistence in the Early Modern Mediterranean*. Baltimore: Johns Hopkins University Press, 2006.

Ebel, Kathryn. "City Views, Imperial Visions: Cartography and the Visual Culture of Urban Space in the Ottoman Empire, 1453–1603." PhD dissertation: University of Texas, Austin, 2002.

———. "Representations of the Frontier in Ottoman Town Views of the Sixteenth Century." *Imago Mundi* 60, no. 1 (2008): 1–22.

Edney, Matthew. "The Irony of Imperial Mapping." In *The Imperial Map: Cartography and the Mastery of Empire*, edited by James Akerman, 11–45. Chicago: University of Chicago Press, 2009.

Eldem, Edhem. "Istanbul: From Imperial to Peripheralized Capital." In *The Ottoman City between East and West: Aleppo, Izmir, and Istanbul*, edited by Edhem Eldem, Daniel Goffman, and Bruce Masters, 135–206. Cambridge: Cambridge University Press, 1999.

Emiralioğlu, Pınar. "Cognizance of the Ottoman World: Visual and Textual Representations in the Sixteenth-Century Ottoman Empire (1514–1596)." PhD dissertation: University of Chicago, 2006.

Ezgü, Fuad. "Piri Reis." In *İslam Ansiklopedisi*, vol. 9, 561–5. Istanbul: MEB, 1964.

Fabris, Antonio. "The Ottoman Mappa Mundi of Hajji Ahmed of Tunis." *Arab Historical Review for Ottoman Studies*, 7–8 (1993): 31–7.

Faroqhi, Suraiya. *The Ottoman Empire and the World Around It*. New York: I. B. Tauris, 2004.

Fernández-Armesto, Felipe. "Maps and Exploration in the Sixteenth and Early Seventeenth Centuries." In *The History of Cartography, Vol. 3, Bk. 1, Cartography in the European Renaissance*, edited by J. B. Harley and David Woodward, 737–58. Chicago: University of Chicago Press, 2007.

———. "Portuguese Expansion in a Global Context." In *Portuguese Oceanic Expansion, 1400–1800*, edited by Francisco Bethencourt and Diogo Ramada Curto, 480–511. Cambridge: Cambridge University Press, 2007.

Fetvacı, Emine. "The Office of Ottoman Court Historian." In *Studies on Istanbul and Beyond: The Freely Papers, Vol. 1*, edited by Robert Ousterhout, 6–21. Philadelphia: University of Pennylvania Museum of Archeology and Anthropology, 2007.

———. *Picturing History at the Ottoman Court*. Bloomington and Indianapolis: Indiana University Press, 2013.

Finlay, Robert. "Prophecy and Politics in Istanbul: Charles V, Sultan Süleyman, and the Habsburg Embassy of 1533–1534." *Journal of Early Modern History* 2, no. 1 (1998): 1–31.

Fleischer, Cornell. *Bureaucrat and Intellectual in the Ottoman Empire: The Historian Mustafa Ali (1541–1600)*. Princeton: Princeton University Press, 1986.

———. "Preliminaries to the Study of the Ottoman Bureaucracy." *Journal of Turkish Studies* 10 (1986): 135–42.

———. "From Şehzade Korkud to Mustafa Ali: Cultural Origins of the Ottoman Nasihatname." In *Third Congress of the Social and Economic History of Turkey, Princeton University, 24–26 August 1983*, edited by Heath Lowry and Ralph S. Hattox, 67–77. Istanbul: ISIS Press, 1990.

———. "The Lawgiver as Messiah: The Making of the Imperial Image in the Reign of Süleyman." In *Soliman le Magnifique et son Temps*, edited by Giles Veinstein, 159–77. Paris: La Documentation Française, 1992.

Fleischer, H. L. "Über das türkische Chatai-name." In *Reprint of Text and Studies on the Historical Geography and Topography of East Asia*, edited by Fuat Sezgin et al., 96–107. Frankfurt: IGAIW, 1993.

Fodor, Pál. "Ottoman Policy towards Hungary." *Acta Orientalia Academiae Scientiarum Hungaricae* 45 (1991): 271–345.

Geary, John S. "Arredondo's *Castillo inexpugnable de la fee*: Anti-Islamic Propaganda in the Age of Charles V." In *Medieval Christian Perceptions of Islam*, edited by John Victor Tolan, 291–311. New York: Garland, 1996.

Gerber, Haim. *Economy and Society in an Ottoman City: Bursa, 1600–1700*. Jerusalem: Hebrew University, 1988.

Goffman, Daniel. *The Ottoman Empire and Early Modern Europe*. New York: Cambridge University Press, 2002.

———. "The Ottoman Empire." In *The Renaissance World*, edited by John Jeffries Martin, 347–64. New York and London: Routledge, 2007.

Gökbilgin, Tayyib. "Kanunî Sultan Süleyman'ın Macaristan ve Avrupa Siyasetinin Sebep ve Âmilleri, Geçirdiği Safhalar." In *Kanuni Armağanı*, 5–41. Ankara: Türk Tarih Kurumu, 1970.

Goodrich, Thomas. "The Search for the Sources of the Sixteenth-Century Tarih-i Hind-i Garbi." *Bulletin of Research in the Humanities* 85 (1982): 269–94.
———. "Atlas-ı Hümayun: A Sixteenth-Century Ottoman Maritime Atlas Discovered in 1984." *Archivum Ottomanicum* 10 (1985): 83–101.
———. "The Earliest Ottoman Maritime Atlas: The Walters Deniz Atlası." *Archivum Ottomanicum* 11 (1986): 25–50.
———. "Some Unpublished Sixteenth-Century Ottoman Maps." In *Comité International d'Études Pré-Ottomanes et Ottomanes, VIth Symposium, Cambridge 1st–4th of July 1984*, edited by J.-L. Bacqué-Grammont and Emeri Van Dunzel, 99–103. Istanbul: IFEA, 1987.
———. *The Ottoman Turks and the New World: A Study of Tarih-i Hind-i Garbi and Sixteenth-Century Ottoman Americana*. Wiesbaden: Harrassowitz, 1990.
———. "Old Maps in the Library of Topkapi Palace in Istanbul." *Imago Mundi* 45 (1993): 120–33.
Goody, Jack. *Renaissances: The One or the Many?* Cambridge: Cambridge University Press, 2010.
Grafton, Anthony. *New Worlds, Ancient Texts: The Power of Tradition and the Shock of Discovery*. Cambridge, MA: Belknap Press, 1992.
Greene, Molly. *A Shared World: Christians and Muslims in the Early Modern Mediterranean*. Princeton: Princeton University Press, 2000.
———. *Catholic Pirates and Greek Merchants: A Maritime History of the Early Modern Mediterranean*. Princeton: Princeton University Press, 2010.
Gruzinski, Serge. *What Time Is It There?*, translated by Jean Birrell. Cambridge and Malden: Polity Press, 2010.
Guilmartin, J. F., Jr. *Gunpowder and Galleys: Changing Technology and Mediterranean Warfare at Sea in the Sixteenth Century*. Cambridge: Cambridge University Press, 1974; London: Marine, 2003.
Hagen, Gottfried. "Katip Çelebi and Tarih-i Hind-i Garbi." *Güney-Doğu Avrupa Araştırmaları Dergisi* 12 (1998): 101–15.
———. *Ein osmanischer Geograph bei der Arbeit Entstehung und Gedankenwelt von Katib Celebis Cihannüma*. Berlin: Klaus Schwarz Verlag, 2003.
———. "Historians of the Ottoman Empire: Katib Çelebi." Originally published on ottomanhistorians.com, 2007. Last accessed via https://www.academia.edu/3488778/Historians_of_the_Ottoman_Empire_Katib_Celebi#
Hamblin, William J. "Arab Perspectives on the Fourth Crusade." In *The Fourth Crusade: Event, Aftermath, and Perceptions* (Papers from the Sixth Conference of the Society for the Study of Crusades and the Latin East, Istanbul, Turkey, 25–29 August, 2004), edited by Thomas F. Madden, 167–79. Burlington, VT: Ashgate, 2008.
Hammer-Purgstall, Joseph von. "Extracts from the Moh'it, that Is the Ocean, A Turkish Work on Navigation in the Indian Seas." *Journal of the Asiatic Society of Bengal* 3, no. 35 (1834): 545–53; 5, no. 56 (1836): 441–68; 6, no. 70 (1837): 805–12; 7, no. 81 (1838): 767–80.
Harley, J. B. "Silences and Secrecy: The Hidden Agenda of Cartography in Early Modern Europe." *Imago Mundi* 40 (1988): 57–76.

Harvey, P. D. A. "Local and Regional Cartography in Medieval Europe." In *The History of Cartography, Vol. 1, Cartography in Prehistoric, Ancient, and Medieval Europe and the Mediterranean*, edited by J. B. Harley and David Woodward, 464–501. Chicago: University of Chicago Press, 1987.

———. *Medieval Maps*. Toronto: University of Toronto Press, 1991.

———. "The Cartographic Context of the Madaba Map." In *The Madaba Map Centenary 1897–1997: Travelling through the Byzantine Umayyad Period*, edited by Michele Piccirillo and Eugenio Alliata, 103–8. Jerusalem: Studium Biblicum Franciscanum, 1999.

Hemmat, Kaveh Louis. "Children of Cain in the Land of Error: A Central Asian Merchant's Treatise on Government and Society in Ming China." *Comparative Studies of South Asia, Africa, and the Middle East* 30, no. 3 (2010): 434–48.

Hess, Andrew C. "The Moriscos: An Ottoman Fifth Column in Sixteenth-Century Spain." *American Historical Review* 74 (1968): 1–25.

———. "The Evolution of the Ottoman Seaborne Empire in the Age of the Oceanic Discoveries 1453–1525." *American Historical Review* 75, no. 7 (1970): 1892–919.

———. "The Ottoman Conquest of Egypt and the Beginning of the Sixteenth-Century World War." *International Journal of Middle Eastern Studies* 4, no. 1 (1973): 55–76.

———. "Piri Reis and the Ottoman Responses to the Voyages of Discovery." *Terrae Incognitae* 6 (1974): 19–37.

———. *The Forgotten Frontier: A History of the Sixteenth-Century Ibero-African Frontier*. Chicago: University of Chicago Press, 1978.

Heywood, Colin. "Between Historical Myth and Mythohistory: The Limits of Ottoman History." *Byzantine and Modern Greek Studies* 12 (1998): 315–45.

———. "The Frontier in Ottoman History: Old Ideas and New Myths." In *Frontiers in Question: Eurasian Borderlands, 700–1700*, edited by Daniel Power and Naomi Standen, 228–51. New York: St. Martin's Press, 1999.

———. "Mehmed II and the Historians: The Reception of Babinger's *Mehmed der Eroberer* during Half a Century." *Turcica* 40 (2008): 295–344.

Hickok, Michael Robert. *Ottoman Military Administration in Eighteenth-Century Bosnia*. Leiden: Brill, 1997.

Hodgson, Marshall G. S. *Rethinking World History: Essays on Europe, Islam and World History*, edited by Edmund Burke. Cambridge: Cambridge University Press, 1993.

Howard, Douglas. "Ottoman Historiography and the Literature of 'Decline' of the Sixteenth and Seventeenth Centuries." *Journal of Asian History* 22, no. 1 (1988): 52–77.

İhsanoğlu, Ekmeleddin. *Osmanlı Coğrafya Literatürü Tarihi*. 2 vols. Istanbul: IRCICA, 2000.

———. "Modern Islam." In *Science and Religion Around the World*, edited by John Hedley Brooke and Ronald L. Numbers, 148–75. Oxford: Oxford University Press, 2012.

Imber, Colin. "Ideals and Legitimation in Early Ottoman History." In *Suleyman the Magnificent and His Age: The Ottoman Empire in the Early Modern World*, edited by Metin Kunt and Christine Woodhead, 138–53. London: Longman, 1995.

———. *Ebu's-Su'ud: The Islamic Legal Tradition*. Palo Alto, CA: Stanford University Press, 1997.

———. *The Ottoman Empire, 1300–1650*. New York: Palgrave Macmillan, 2002.

———. "Frozen Legitimacy." In *Legitimizing the Order: The Ottoman Rhetoric of State Power*, edited by Hakan Karateke, 99–111. Leiden: Brill, 2005.

İnalcık, Halil. "Bursa." *EI²*.

———. "Selim I." *EI²*.

———. "Mehmed the Conqueror (1432–1481) and His Time." *Speculum* 25 (1960): 408–27.

———. "The Rise of Ottoman Historiography." In *Historical Writing on the Peoples of Asia: Historians of the Middle East*, edited by Bernard Lewis and P. M. Holt, 152–67. London: Oxford University Press, 1962.

———. "The Policy of Mehmed II toward the Greek Population of Istanbul and the Byzantine Buildings of the City." *Dumbarton Oak Papers* 23 (1970): 231–49.

———. *The Ottoman Empire: The Classical Age 1300–1600*. New Rochelle: Orpheus Publishing, 1973.

———. "The Hub of the City: The Bedestan of Istanbul." *International Journal of Turkish Studies* 1 (1980): 1–17.

———. "Military and Fiscal Transformation in the Ottoman Empire." *Archivum Ottomanicum* 6 (1980): 283–337.

———. *Fatih Devri Üzerinde Tetkikler ve Vesikalar*. Ankara: Türk Tarih Kurumu, 1995.

İnalcık, Halil, and Donald Quataert, eds. *An Economic and Social History of the Ottoman Empire, Vol. 1: The Ottoman State, Economy, and Society 1300–1600*. Cambridge: Cambridge University Press, 1994.

Isom-Verhaaren, Christine. "'Barbarossa and His Army Who Came to Succor All of Us': Ottoman and French Views of Their Joint Campaign of 1543–1544." *French Historical Studies* 30, no. 3 (2007): 395–425.

Itzkowitz, Norman. *Ottoman Empire and Islamic Tradition*. Chicago: University of Chicago Press, 1972.

Jardine, Lisa, and Jerry Brotton. *Global Interests: Renaissance Art between East and West*. Ithaca: Cornell University Press, 2000.

John, Jeremy, and Emilie Savage-Smith. "The Book of Curiosities: A Newly Discovered Series of Islamic Maps." *Imago Mundi* 55 (2003): 7–24.

Johnston, Norman J. "The Urban World of the Matraki Manuscript." *Near Eastern Studies Journal* 30, no. 3 (1971): 159–76.

Kafadar, Cemal. "When Coins Turned into Drops of Dew and Bankers Became Robbers of Shadows: The Boundaries of Ottoman Economic Imagination at the End of the Sixteenth Century." PhD dissertation: McGill University, 1986.

———. "The Myth of the Golden Age: Ottoman Historical Consciousness in the Post-Süleymanic Era." In *Süleyman the Second and His Time*, edited by Halil Inalcik and Cemal Kafadar, 37–48. Istanbul: Isis Press, 1993.

———. "The Ottomans and Europe." In *Structures and Assertions*. Vol. 1 of *Handbook of European History 1400–1600: Late Middle Ages, Renaissance and Reformation*, edited by Thomas A. Brady, Jr., Heiko A. Oberman, and James D. Tracy. Leiden: E. J. Brill, 1994.

———. *Between Two Worlds: The Construction of the Ottoman State*. Berkeley and Los Angeles: University of California Press, 1996.

———. "The Question of Ottoman Decline." *Harvard Middle Eastern and Islamic Review* 4, no. 1–2 (1997–98): 30–76.

Kafescioğlu, Çiğdem. *Constantinopolis/Istanbul: Cultural Encounter, Imperial Vision, and the Construction of the Ottoman Capital*. University Park: Penn State University Press, 2009.

Kagan, Richard and Benjamin Schmidt. "Maps and the Early Modern State: Official Cartography." In *The History of Cartography, Vol. 3, Bk. 1, Cartography in the European Renaissance*, edited by J. B. Harley and David Woodward, 661–79. Chicago: University of Chicago Press, 2007.

Kahlaoui, Tarek. "The Depiction of the Mediterranean in Islamic Cartography (11th–16th Centuries): The *Suras* (Images) of the Mediterranean from the Bureaucrats to the Sea Captains." PhD dissertation: University of Pennsylvania, 2008.

Kahle, Paul. "Piri Re'is: The Turkish Sailor and Cartographer." *Journal of the Pakistan Historical Society* 4 (1956): 99–108.

———. "Eine Islamische Quelle über China um 1500 (Das Khitayname des Ali Ekber)." In *Reprint of Text and Studies on the Historical Geography and Topography of East Asia*, edited by Fuat Sezgin et al., 379–98. Frankfurt: IGAIW, 1993.

Kaptein, Laban. *Eindtijd en Antichrist (Ad-Dağğal) in de Islam: Eschatologie bij Ahmet Bican († ca. 1466)*. Leiden: CNWS, 1997.

Karamustafa, Ahmet. "Cosmographical Diagrams." In *The History of Cartography, Vol. 2, Bk. 1, Cartography in the Traditional Islamic and South Asian Societies*, edited by J. B. Harley and David Woodward, 71–89. Chicago: University of Chicago Press, 1992.

———. "Introduction to Ottoman Cartography." In *The History of Cartography, Vol. 2, Bk. 1, Cartography in the Traditional Islamic and South Asian Societies*, edited by J. B. Harley and David Woodward, 205–8. Chicago: University of Chicago Press, 1992.

———. "Military, Administrative, and Scholarly Maps and Plans." In *The History of Cartography, Vol. 2, Bk. 1, Cartography in the Traditional Islamic and South Asian Societies*, edited by J. B. Harley and David Woodward, 209–28. Chicago: University of Chicago Press, 1992.

Karrow, Robert. *Mapmakers of the Sixteenth Century and Their Maps: Bio-Bibliographies of the Cartographers of Abraham Ortelius, 1570*. Chicago: Published for the Newberry by Speculum Orbis Press, 1993.

Keller, Marcus. "Nicolas de Nicolay's *Navigations* and the Domestic Politics of Travel Writing." *L'esprit créateur* 48, no. 1 (2008): 18–31.

Kelly, Gavin. "The New Rome and the Old: Ammianus Marcellinus' Silences on Constantinople." *Classical Quarterly* 53, no. 2 (2003): 588–607.

King, David A. "Taki al-Din." *EI²*.

King, David A., and Richard P. Lorch. "Qibla Charts, Qibla Maps, and Related Instruments." In *The History of Cartography, Vol. 2, Bk. 1, Cartography in the Traditional Islamic and South Asian Societies*, edited by J. B. Harley and David Woodward, 189–203. Chicago: University of Chicago Press, 1992.

Kırzıoğlu, M. Fahrettin. *Osmanlılar'ın Kafkas-Ellerini Fethi (1451–1590)*. Ankara: Sevinç Matbaası, 1976.

Kish, George. *The Suppressed Turkish Map of 1560*. Ann Arbor: William L. Clements Library, 1957.

Kissling, Hans Joachim. *Der See-Atlas des Sejjid Nuh*. Munich: Rudolf Trofenik, 1966.

Kivelson, Valerie. *Cartographies of Tsardom: The Land and Its Meanings in Seventeenth-Century Russia*. Ithaca: Cornell University Press, 2006.

Kleinschmidt, Harald. *Charles V: The World Emperor*. Phoenix Mill, UK: Sutton Publishing, 2004.

Kreiser, Klaus. "Piri Reis." In *Lexicon zur Geschichte der Kartographie*, edited by Ingrid Kretschmer, Johannes Dörflinger, and Franz Wawrik, vol. 2, 607–9. Vienna: Franz Deuticke, 1986.

Krstić, Tijana. "Illuminated by the Light of Islam and the Glory of the Ottoman Sultanate: Self-Narratives of Conversion to Islam in the Age of Confessionalization." *Comparative Studies in Society and History* 51, no. 1 (2009): 35–63.

Kuban, Doğan. *İstanbul Bir Kent Tarihi: Bizantion, Konstantinopolis, İstanbul*, translated by Zeynep Rona. Istanbul: Türkiye Ekonomik ve Toplumsal Tarih Vakfı, 1996.

Kurtoğlu, Fevzi. "Amiral Selman Reis Layihasi." *Deniz Mecmuasi* 47, no. 335 (1935): 67–73.

Kut, Günay. "Matba'a." *EI²*.

———. *Acaibü'l-Mahlukat: Eski Türk Edebiyatı Araştırmaları II*, edited by Fatma Büyükkarcı Yılmaz. Istanbul: Simurg Kitapçılık, 2010.

LeDonne, John P. *The Grand Strategy of the Russian Empire, 1650–1831*. Oxford: Oxford University Press, 2004.

Lestringant, Frank. *Mapping the Renaissance World: The Geographical Imagination in the Age of Discovery*, translated by David Fausett. Cambridge: Polity Press, 1994.

Lestringant, Frank, and Monique Pelletier. "Maps and Descriptions of the World in Sixteenth-Century France." In *The History of Cartography, Vol. 3, Bk. 2, Cartography in the European Renaissance*, edited by J. B. Harley and David Woodward, 1463–79. Chicago: University of Chicago Press, 2007.

Lewis, Bernard. "Ottoman Observers of Ottoman Decline." *Islamic Studies* 1 (1962): 71–87.

———. *The Emergence of Modern Turkey*. Oxford: Oxford University Press, 1968.

Liu, Yih-Min. "A Comparative and Critical Study of Ali Akbar's Khitay-nama with Reference to Chinese Sources." *Central Asiatic Journal* 27 (1983): 58–78.

Livingstone, David. *The Geographical Tradition: Episodes in the History of a Contested Enterprise*. Oxford: Blackwell Publishing, 1992.

Lowry, Heath W. "The Ottoman Tahrir Defterleri as a Source for Social and Economic History: Pitfalls and Limitations." In *Studies in Defterology: Ottoman Society in the Fifteenth and Sixteenth Centuries*, edited by Heath W. Lowry, 3–18. Istanbul: ISIS Press, 1992.

Lozovsky, Natalia. *"The Earth Is Our Book": Geographical Knowledge in the Latin West ca. 400–1000*. Ann Arbor: University of Michigan Press, 2000.

MacLean, Gerald. *Re-Orienting the Renaissance: Cultural Exchanges with the East*. New York: Palgrave Macmillan, 2005.

Mangani, Giorgio. "Abraham Ortelius and the Hermetic Meaning of the Cordiform Projection." *Imago Mundi* 50 (1998): 59–82.

Manners, Ian. "Constructing the Image of a City: The Representation of Constantinople in Christopher Buondelmonti's *Liber Insularum Archipelagi*." *Annals of the American Association of Geography* 87 (1997): 72–102.

———. *European Cartographers and the Ottoman World: Maps from the Collection of O.J. Sopranos*. Chicago: Oriental Institute Museum of the University of Chicago, 2007.

Martin, John Jeffries. "The Renaissance: A World in Motion." In *The Renaissance World*, edited by John Jeffries Martin, 3–31. New York and London: Routledge, 2007.

Matuz, Joseph, ed. *L'Ouvrage de Seyfi Çelebi: Historien Ottoman du XVIe Siecle*. Paris: Dépositaire, 1968.

McConnell, Anita. "L. F. Marsigli's Voyage to London and Holland, 1721–1722." *Notes and Records of the Royal Society of London* 41, no. 1 (1986): 39–76.

Ménage, V. L. "The Map of Hajji Ahmed and Its Makers." *BSOAS* 21 (1958): 299–311.

———. "The Beginnings of Ottoman Historiography." In *Historical Writing on the Peoples of Asia: Historians of the Middle East*, edited by Bernard Lewis and P. M. Holt, 168–79. London: Oxford University Press, 1962.

Meserve, Margaret. *Empires of Islam in Renaissance Historical Thought*. Cambridge, MA: Harvard University Press, 2008.

Mikhail, Alan, and Christine M. Philliou. "The Ottoman Empire and the Imperial Turn." *Society for the Comparative Study of Society and History* 54, no. 4 (2012): 721–45.

Minorsky, V. "The Poetry of Shah İsma'il I." *BSOAS* 10 (1940–42): 1006a–1053a.

Miquel, Andre. "al-Istakhri, abu Ishak İbrahim b. Muhammad al-Farisi al-Karkhi." *EI²*.

———. "Iqlim." *EI²*.

Mitchel, Colin P. *The Practice of Politics: Power, Religion, and Rhetoric*. London and New York: I. B. Tauris, 2009.

Monfasani, John. *George of Trebizond: A Biography and a Study of His Rhetoric and Logic*. Leiden: Brill, 1976.

Mordtmann, J. H. "Das Observatorium des Taqi ed-Din zu Pera." In *Astronomische Instrumente Publikationen 1918–1925*. Vol. 4 of *Arabische Instrumente in orientalischen Studien*, edited by Fuat Sezgin, 281–95. Frankfurt: IGAIW, 1991.

Morris, M. "Miroir des pays, ou relation des voyages de Sidi Aby fils d'Housain, nommé ordinairement Katibi Roumi, Amiral de Soliman II." *Journal Asiatique* 9 (1826): 27–286.

Mughul, Yakub. *Kanuni Devri Osmanlılar'ın Hint Okyanusu Politikası ve Osmanlı-Hint Müslümanları Münasebetleri 1517–1538*. Istanbul: Fetih Yayınevi, 1974.

Mundy, Barbara E. *The Mapping of New Spain: Indigenous Cartography and the Maps of the Relaciones Geograficas*. Chicago: University of Chicago Press, 1996.

Murphy, Rhoads. "Seyyid Muradi's Prose Biography of Hizir Ibn Yakub, alias Hayreddin Barbarossa." *Acta Orientalia Academiae Scientiarum* 54, no. 4 (2001): 519–32.

Necipoğlu, Gülru. "The Süleymaniye Complex in Istanbul: An Interpretation." *Muqarnas* 3 (1985): 92–117.

———. "Süleyman the Magnificent and the Representation of Power in the Context of Ottoman-Hapsburg-Papal Rivalry." *Art Bulletin* 71, no. 3 (1989): 401–27.

———. *Architecture, Ceremonial, and Power: The Topkapı Palace in the Fifteenth and Sixteenth Centuries*. New York: Architectural History Foundation, 1991.

———. "A Kanun for the State, a Canon for the Arts." In *Soliman le Magnifique et son Temps*, edited by Giles Veinstein, 195–216. Paris: La Documentation Française, 1992.

———. "Challenging the Past: Sinan and the Competitive Discourse of Early Modern Islamic Architecture." *Muqarnas* 10 (1993): 169–80.

———. *The Age of Sinan: Architectural Culture in the Ottoman Empire*. Princeton: Princeton University Press, 2005.

Nelson, Janet. "Kingship and Empire." In *The Cambridge History of Medieval Thought c. 350–c. 1450*, edited by J. H. Burns, 211–52. Cambridge: Cambridge University Press, 1988.

Newman, Andrew. *Safavid Iran: Rebirth of a Persian Empire*. London: I. B. Tauris, 2006.

Nicol, D. M. "Byzantine Political Thought." In *The Cambridge History of Medieval Thought c. 350–c. 1450*, edited by J. H. Burns, 51–83. Cambridge: Cambridge University Press, 1988.

Nussbaum, Felicity A., ed. *The Global Eighteenth Century*. Baltimore: Johns Hopkins University Press, 2003.

Ocak, Ahmet Yaşar. *Osmanlı Toplumunda Zındıklar ve Mulhidler (15.-17. Yüzyıllar)*. Istanbul: Tarih Vakfi Yurt Yayinlari, 1998.

Orbay, İffet. "Istanbul Viewed: The Representation of the City in Ottoman Maps of the Sixteenth and Seventeenth Centuries." PhD dissertation: MIT, 2001.

Orhonlu, Cengiz. "Hadım Süleyman Pasha," EI^2.

———. "Hint Kaptanlığı ve Piri Reis," *Belleten* 34, no. 134 (1967): 235–54.

———. "Seydi Ali Reis." *Tarih Enstitüsü Dergisi* 1 (1970): 39–56.

———. *Osmanlı İmparatorluğu'nun Güney Siyaseti: Habeş Eyaleti*. Istanbul: TTK, 1974.

Özbaran, Salih. "Osmanlı İmparatorluğu ve Hindistan Yolu: Onaltıncı Yüzyılda Ticaret Yolları Üzerinde Türk-Portekiz Rekabet ve İlişkileri." *Tarih Dergisi* 31 (1977): 66–146.

———. "A Turkish Report in the Red Sea and the Portuguese in the Indian Ocean (1525)." *Arabian Studies* 4 (1978): 81–8.

———. "Expansion in the Southern Seas." In *Süleyman the Second and His Time*, edited by Halil İnalcık and Cemal Kafadar, 241–9. Istanbul: ISIS Press, 1993.

———. *The Ottoman Response to European Expansion: Studies on Ottoman-Portuguese Relations in the Indian Ocean and Ottoman Administration in the Arab Lands during the Sixteenth Century*. Istanbul: ISIS Press, 1994.

———. *Ottoman Expansion towards the Indian Ocean in the 16th Century*. Istanbul: Bilgi University, 2009.

Özdemir, Kemal. *Ottoman Nautical Charts: The Atlas of Ali Macar Reis*, translated by P. Mary Işın. Istanbul: Marmara Bank Publication, 1992.

Özen, Mine Esiner. *Piri Reis and His Charts*. Istanbul: N. Refioğlu, 1998.

Özendeş, Engin. *The First Ottoman Capital: Bursa*. Istanbul: Yapı Endüstri Merkezi Yay., 1999.

Özkan, Behlül. *From the Abode of Islam to the Turkish Vatan: The Making of a National Homeland in Turkey*. New Haven and London: Yale University Press, 2012.

Padrón, Ricardo. *The Spacious Word: Cartography, Literature, and Empire in Early Modern Spain*. Chicago: University of Chicago Press, 2004.

Pagden, Anthony. *Lords of All the World: Ideologies of Empire in Spain, Britain and France c. 1500–c. 1800*. New Haven: Yale University Press, 1995.

Parker, Geoffrey. *The Grand Strategy of Philip II*. New Haven: Yale University Press, 1998.

Pearson, Michael N. *Merchants and Rulers in Gujarat: The Response to the Portuguese in the Sixteenth Century*. Berkeley and Los Angeles: University of California Press, 1976.

———. "The Search for Similar: Early Contacts between Portuguese and Indians." In *Clash of Cultures: Essays in Honour of Niels Steensgaard*, edited by Jens Christian V. Johansen, Erling Ladewig Petersen, and Henrik Stevensborg, 144–59. Odense: Odense University Press, 1992.

Peirce, Leslie P. *The Imperial Harem: Women and Sovereignty in the Ottoman Empire*. Oxford: Oxford University Press, 1993.

Pinto, Karen C. "Ways of Seeing: Scenarios of the World in the Medieval Islamic Cartographic Imagination." PhD dissertation: Columbia University, 2001.

Portuondo, Maria M. *Secret Science: Spanish Cosmography and the New World*. Chicago: University of Chicago Press, 2009.

Quataert, Donald. *The Ottoman Empire, 1700–1922*. Cambridge: Cambridge University Press, 2005.

Reeves, Marjorie. "The Medieval Heritage." In *Prophetic Rome in the Renaissance Period*, edited by Marjorie Reeves, 3–21. Oxford: Clarendon Press, 1992.

Roberts, Sean. "Cartography Between Cultures: Francesco Berlinghieri's 'Geographia' of 1482." PhD dissertation: University of Michigan, 2006.

———. *Printing a Mediterranean World: Florence, Constantinople, and the Renaissance Geography*. Cambridge, MA: Harvard University Press, 2013.

Rogers, J. M. "Itineraries and Town Views in Ottoman Histories." In *The History of Cartography, Vol. 2, Bk. 2,Cartography in the Traditional Islamic and South Asian Societies*, edited by J. B. Harley and David Woodward, 228–56. Chicago: University of Chicago Press, 1992.

Rothman, Natalie. "Becoming Venetian: Conversion and Transformation in the Seventeenth-Century Mediterranean," *Mediterranean Historical Review* 21, no. 1 (2006): 39–75.

———. *Brokering Empire: Trans-Imperial Subjects Between Venice and Istanbul*. Ithaca: Cornell University Press, 2011.

Sabev, Orlin. "The First Ottoman Turkish Printing Enterprise: Success or Failure." In *Ottoman Tulips, Ottoman Coffee: Leisure and Lifestyle in the Eighteenth Century*, edited by Dana Sajdi, 63–95. London: I. B. Tauris, 2007.

Safier, Neil. "The Confines of the Colony: Boundaries, Ethnographic Landscapes, and Imperial Cartography in Iberoamerica." In *The Imperial Map: Cartography and the Mastery of Empire*, edited by James Akerman, 133–85. Chicago: University of Chicago Press, 2009.

Saliba, George. "The Role of the Astrologer in Medieval Islamic Society." *Bulletin D'Études Orientales* 44 (1992): 45–67.

Sandman, Alison. "Spanish Nautical Geography in the Renaissance." In *The History of Cartography, Vol. 3, Bk. 1, Cartography in the European Renaissance*, edited by J. B. Harley and David Woodward, 1095–171. Chicago: University of Chicago Press, 2007.

Sayılı, Adnan. *The Observatory in Islam and Its Place in the General History of the Observatory*. Ankara: Türk Tarih Kurumu, 1960.

Schefer, Charles. "Trois Chapitres du Khitay Nameh: Texte Persan et Traduction Française." In *Reprint of Text and Studies on the Historical Geography and Topography of East Asia*, edited by Fuat Sezgin et al., 159–214. Frankfurt: IGAIW, 1993.

Selen, Hamit Sadi. "16. Asırda Yapılmış Anadolu Atlası." In *II. Türk Tarih Kongresi, Istanbul, 20–25 Eylül 1937, Kongreye Sunulan Bildiriler*, 813–17. Istanbul: Kenan Matbaasi, 1943.

Setton, Kenneth M. *Europe and the Levant in the Middle Ages and the Renaissance*. London: Variorum Reprints, 1974.

———. *The Papacy and the Levant (1204–1571)*. 4 vols. Philadelphia: American Philosophical Society, 1984.

———. *Western Hostility to Islam and Prophecies of Turkish Doom*. Philadelphia: American Philosophical Society, 1992.

Sezgin, Fuat. *Geschichte des Arabischen Schrifttums, Band X, Mathematische Geographie und Kartographie im Islam und ihr Fortleben im Abendland*. 3 vols. Frankfurt am Main: Institut für Geschichte der Arabisch-Islamischen Wissenschaften an der Johann Wolfgang Goethe-Universität, 2000.

Sezgin, Fuat, et al., eds. *Studies on the Works of Sidi Ali*. Vol. 16 of *Islamic Geography*. Frankfurt: IGAIW, 1992.

Sheehan, Kevin Joseph. "Iberian Asia: The Strategies of Spanish and Portuguese Empire Building, 1540–1700." PhD dissertation: University of California, Berkeley, 2008.

Simek, Rudolph. *Heaven and Earth in the Middle Ages: The Physical World Before Columbus*, translated by Angela Hall. Woodbridge, UK: Boydell & Brewer, 1996.

Soucek, Svat. "Piri Reis." *EI²*.

———. "Sidi Ali Reis." *EI²*.

———. "The 'Ali Macar Reis Atlas' and the Deniz Kitabı: Their Place in the Genre of Portolan Charts and Atlases." *Imago Mundi* 25 (1971): 17–27.

———. "The Rise of the Barbarossas in North Africa." *Archivum Ottomanicum* 3 (1972): 228–50.

———. "Islamic Charting in the Mediterranean." In *The History of Cartography, Vol. 2, Bk. 1, Cartography in the Traditional Islamic and South Asian Societies*, edited by J. B. Harley and David Woodward, 263–87. Chicago: University of Chicago Press, 1992.

———. "Piri Reis." In *Süleyman the Second and His Time*, edited by Halil İnalcık and Cemal Kafadar, 379–89. Istanbul: ISIS Press, 1993.

———. "Piri Reis and the Ottoman Discovery of the Great Discoveries." *Studia Islamica* 79 (1994): 121–42.

———. *Piri Reis and Turkish Mapmaking after Columbus*. Oxford: Oxford University Press, 1996.

Stoye, John. *Marsigli's Europe 1680–1730: The Life and Times of Luigi Ferdinando Marsigli, Soldier and Virtuoso*. New Haven and London: Yale University Press, 1994.

Subrahmanyam, Sanjay. "Connected Histories: Notes towards a Reconfiguration of Early Modern Eurasia." *Modern Asian Studies* 31 (1997): 735–62.

———. "A Note on the Rise of Surat in the Sixteenth Century." *JESHO* 43, no. 1 (2000): 23–33.

Taeschner, Franz. "Die Geographische Literatur der Osmanen." *ZDMG* 77, no. 2 (1923): 31–80.

———. "Osmanlilarda Coğrafya." *Türkiyat Mecmuasi* 2 (1926): 271–314.

———. "Zur Geschichte des Djihannuma." *Mitteilungen des Seminars für Orientalische Sprachen* 2, no. 29 (1926): 99–110.

———. "Djughrafiya: The Ottoman Geographers." *EI²*.

———. "The Itinerary of the First Persian Campaign of Sultan Süleyman, 1534–36, According to Nasuh al-Matraki." *Imago Mundi* 13 (1956): 53–5.

Tansel, Selahattin. *Yavuz Sultan Selim*. Ankara: Milli Eğitim Basımevi, 1969.

Tekindağ, Şahabettin. "Şah Kulu Baba Tekeli İsyanı." *Belgelerle Türk Tarihi Dergisi* 3 (1967/68): 34–9; 4 (1967/68): 54–9.

Terzioğlu, Derin. "The Imperial Circumcision Festival of 1582: An Interpretation." *Muqarnas* 12 (1995): 84–100.

Tezcan, Baki. "The Multiple Faces of the One: The Invocation Section of Ottoman Literary Introductions as a Locus for the Central Argument of the Text." *Middle Eastern Literatures* 12 (2009): 27–41.

———. *The Second Ottoman Empire: Political and Social Transformation in the Early Modern World.* Cambridge: Cambridge University Press, 2010.

———. "The 'Frank' in the Ottoman Eye of 1583." In *The Turk and Islam in the Western Eye (1453–1750): Visual Imagery before Orientalism*, edited by James Harper, 267–96. Burlington, VT: Ashgate, 2011.

Thrower, Norman. *Maps and Civilization: Cartography in Culture and Society.* 2nd ed. Chicago: University of Chicago Press, 1999.

Tibbets, Gerald R. "The Balkhi School of Geographers." In *The History of Cartography, Vol. 2, Bk. 1, Cartography in the Traditional Islamic and South Asian Societies*, edited by J. B. Harley and David Woodward, 108–36. Chicago: University of Chicago Press, 1992.

Tolias, George. "Isolarii, Fifteenth to Seventeenth Century." In *The History of Cartography, Vol. 3, Bk. 1, Cartography in the European Renaissance*, edited by J. B. Harley and David Woodward, 263–85. Chicago: University of Chicago Press, 2007.

Török, Zsolt. "Luigi Ferdinando Marsigli (1658–1730) and Early Thematic Mapping in the History of Cartography." In *Térkép – tudomány / Map – Science*, edited by L. Zentai, J. Györffy, and Zs. Török, 403–13. Budapest: ELTE Térképtudományi és Geoinformatikai Tanszék, 2006.

Turan, Ebru. "The Sultan's Favorite: Ibrahim Pasha and the Making of the Ottoman Universal Sovereignty in the Reign of Sultan Süleyman (1516–1526)." PhD dissertation: University of Chicago, 2007.

Türkay, Cevdet. *Osmanlı Türklerinde Coğrafya.* Istanbul: Maarif Basımevi, 1959.

Turnbull, David. "Cartography and Science in Early Modern Europe: Mapping the Construction of Knowledge Spaces." *Imago Mundi* 48 (1996): 5–24.

Uğur, Ahmed. *The Reign of Selim I in the Light of the Selim-name Literature.* Berlin: K. Schwarz, 1985.

Uğur, Yunus. "Bursa." In *Encyclopedia of the Ottoman Empire*, edited by Gabor Agoston and Bruce Masters, 105–6. New York: Facts on File, 2009.

Uluçay, Çağatay. "Yavuz Sultan Selim Nasıl Padişah Oldu?" *Tarih Dergisi* 9 (1953): 53–90; 10 (1954): 117–42; 11 (1955): 187–200.

Unat, Yavuz. "Mustafa İbn Ali el-Muvakkit ve İ'lam el-'İbad fi A'lam el-Bilad (Şehirler Aleminde Mesafelerin Bildirimi) Adlı Risalesi." *EJOS* 7, no. 10 (2004): 1–47.

Uzunçarşılı, *İsmail Hakkı. Osmanlı Tarihi.* 4 vols. Ankara: Türk Tarih Kurumu, 1947–59.

———. *Osmanlı Devletinin Saray Teşkilatı.* Ankara: TTK, 1988.

Vatin, Nicolas. "Pourquoi un Turc Ottoman Racontait-il son Voyage? Note sur Les Ottomans des Vaki'at-i Sultan Cem au Seyahatname d'Evliya Çelebi." In *Les Ottomans et L'Occident (XVe–XVIe siècles)*, 179–93. Istanbul: Les Éditions ISIS, 2001.

———. *Rodos* Şövalyeleri ve Osmanlılar: *Doğu Akdeniz'de savaş diplomasi ve korsanlık: 1480–1522*, translated by Tülin Altınova. Istanbul: Türkiye Ekonomik ve Toplumsal Tarih Vakfı, 2004.

Vedovato, Mirco. "The Nautical Chart of Mohammed Raus, 1590." *Imago Mundi* 8 (1951): 49.

Veres, Madalina-Valeria. "Redefining Imperial Borders: Marking the Eastern Border of the Habsburg Monarchy in the Second Half of the Eighteenth Century." In *History of Cartography, Lecture Notes in Geoinformation and Cartography*, edited by E. Liebenberg, P. Collier, and Zs. Török. Heidelberg: Springer-Verlag, 2014.

Vlahakis, George. *Imperialism and Science: Social Impact and Interaction*. Santa Barbara, CA: ABC-CLIO, 2006.

Vogel, Klaus A. "Cosmography." In *The Cambridge History of Science, Vol. 3, Early Modern Science*, edited by Lorraine Daston and Katherine Park, 469–97. Cambridge: Cambridge University Press, 2006.

Wallerstein, Immanuel. *The Modern World System I: Capitalist Agriculture and the Origins of the European World-Economy in the Sixteenth Century*. New York and London: Academic Press, 1974.

White, Sam. *The Climate of Rebellion in the Early Modern Ottoman Empire*. Cambridge: Cambridge University Press, 2011.

Withers, Charles. *Placing the Enlightenment: Thinking Geographically about the Age of Reason*. Chicago: University of Chicago, 2007.

Withers, Charles, and Robert J. Mayhew. "Geography: Space, Place and Intellectual History in the Eighteenth Century." *Journal for Eighteenth-Century Studies* 34, no. 4 (2011): 445–52.

Woodhead, Christine. "Rüstem Pasha," *EI*[2].

———. "Seyfi." *EI*[2].

———. "An Experiment in Official Historiography: The Post of Official Şehnameci in the Ottoman Empire, c. 1555–1605." *Wiener Zeitschrift fur die Kundedes Morgenlandes* 75 (1983): 157–82.

———. "Murad III and the Historians of Ottoman Imperial Authority in Late 16th-Century Historiography." In *Legitimizing the Order: The Ottoman Rhetoric of State Power*, edited by Hakan Karateke, 85–98. Leiden: Brill, 2005

Woodward, David. "Medieval Mappaemundi." In *The History of Cartography, Vol. 1, Cartography in Prehistoric, Ancient, and Medieval Europe and the Mediterranean*, edited by J. B. Harley and David Woodward, 286–370. Chicago: University of Chicago Press, 1987.

———. "Cartography and the Renaissance: Continuity and Change." In *The History of Cartography, Vol. 3, Bk. 1, Cartography in the European Renaissance*, edited by J. B. Harley and David Woodward, 3–24. Chicago: University of Chicago Press, 2007.

Yates, Frances A. *Astrea: The Imperial Theme in the Sixteenth Century*. London: Ark, 1985.

Yavuz, Hulusi. *Kabe ve Haremeyn İçin Yemen'de Osmanli Hakimiyeti*. Istanbul: Serbest Matbaası, 1984.

Yerasimos, Stéphane. *Légendes d'empire: La fondation de Constantinople et de Sainte-Sophie dans les traditions turques*. Paris: Librairie d'Amérique et d'Orient J. Maisonneuve, 1990.

———. *Les Voyageurs Dans L'Empire Ottoman (XIVe–XVIe siècles): Bibliographie, Itinéraires et Inventaire des Lieux Habités*. Ankara: TTK, 1991.

———. *Constantinople: Istanbul's Historical Heritage*. Paris: Konemann, 2005.

Yih-Min, Lin. *Ali Ekber'in Hitayname Adlı Eserinin Çin Kaynakları ile Mukayese ve Tenkidi*. Taipei, 1967.

———. "A Comparative and Critical Study of Ali Ekber's Khitay-nama with Reference to Chinese Sources." *Central Asiatic Journal* 27 (1983): 58–78.

Young, Soo. "Minding the Shop: Zati and the Making of Ottoman Poetry in the First Half of the Sixteenth Century." PhD dissertation: University of Chicago, 2005.

Yurdaydın, Hüseyin. *Matrakçı Nasuh*. Ankara: Ankara University Press, 1963.

———. "Muradi ve Eserleri." *Belleten* 27 (1963): 453–66.

Zenker, J. Th. "Das chinesische Reich, nach dem türkischen Khatainame." In *Reprint of Text and Studies on the Historical Geography and Topography of East Asia*, edited by Fuat Sezgin et al., 109–29. Frankfurt: IGAIW, 1993.

Index

Note: Page illustrations in **bold type** indicate illustrations.

For Product Safety Concerns and Information please contact our EU representative GPSR@taylorandfrancis.com Taylor & Francis Verlag GmbH, Kaufingerstraße 24, 80331 München, Germany